Geography and Disasters

Geography and Disasters

Places, Processes and the Human Geographical Imagination

Edited by
Nathaniel O'Grady and Gemma Sou

BLOOMSBURY ACADEMIC
LONDON • NEW YORK • OXFORD • NEW DELHI • SYDNEY

BLOOMSBURY ACADEMIC

Bloomsbury Publishing Plc,
50 Bedford Square,
London,
WC1B 3DP,
UK

Bloomsbury Publishing Inc,
1359 Broadway,
New York,
NY 10018,
USA

Bloomsbury Publishing Ireland, 29 Earlsfort Terrace, Dublin 2, D02 AY28, Ireland

BLOOMSBURY, BLOOMSBURY ACADEMIC and the Diana logo are trademarks of Bloomsbury Publishing Plc

First published in the United States of America 2026

Library of Congress Cataloging-in-Publication Data Available

ISBN: HB: 978-1-6669-7088-3
ePDF: 978-1-6669-7089-0
eBook: 978-1-9787-6904-5

Typeset by Amnet

Contents

List of Figures

List of Tables

Notes on Contributors

Amitangshu Acharya is Lecturer in water governance at the IHE Delft Institute for Water Education, the Netherlands. A human geographer by training, he specializes in the interdisciplinary field of political ecology, focusing on postcolonial nature in South Asia, where he explores the technomediation of ruptured relationships between people and their environment. He completed his PhD at the University of Edinburgh as a Leverhulme Trust scholar, and his thesis investigated the politics of drinking water quality and urbanization in Bhuj, a small city in the western Indian state of Gujarat. Amitangshu coordinates the decolonizing science PhD course in the Netherlands and has written on water cultures, the climate crisis and everyday urbanism in *The Economist*, *The Independent*, *Huffington Post*, *The Hindu*, *Khaleej Times*, *Indian Express*, *Scroll.in* and *The Mint*.

Teresa Armijos Burneo is Senior Lecturer in environmental sustainability at the School of Geosciences, University of Edinburgh. She conducts interdisciplinary and decolonial research that combines critical social theory and arts-based participatory methods in disaster risk studies. Her research aims to co-produce knowledge with urban, rural and indigenous communities, supporting them to reduce risk and recover from disasters in different countries across Latin America and the Caribbean.

Belén Desmaison is Lecturer and Researcher at the Pontificia Universidad Católica del Perú (PUCP, Peru). Her areas of action research include citizen participation, climate change, Amazonia and urbanism. She holds a BA in architecture from the University of Pennsylvania (USA), a bachelor of architecture from PUCP (Peru) and an MSc in building and urban design in development from the Bartlett – Development Planning Unit, University College London (UK), and is a PhD candidate in human geography at Durham University (UK) and coordinator of the action-research projects Ciudades Auto-Sostenibles Amazónicas (CASA), Knowledge in Action Towards Urban Equality (Lima) (KNOW), Gender Responsive Resilience and Intersectionality in Policy and Practice (GRRIPP) and Bote Auto-suficiente para Laboratorio y Servicios en la Amazonía (BALSA). Belén has received recognition at the World Architecture Forum (2018), the

Peruvian Biennial of Architecture and Urbanism (2019 and 2023), the Ibero-American Biennial of Architecture and Urbanism (2022) and the International Union of Architects (2023).

Nimesh Dhungana is Lecturer in disasters and global health at the Humanitarian and Conflict Response Institute, University of Manchester. He is an interdisciplinary scholar of development and disaster studies whose research and teaching explore the relationship between disasters and democratic politics, with particular empirical emphasis on participatory and grassroots responses to disasters. He completed his PhD in 2019 at the London School of Economics and Political Science (LSE). Previously, Nimesh worked as a research manager at the Center on Health Risk and Society, the American University, United States. Before entering academia, he worked in the development sector in Nepal, leading development evaluations and research while collaborating with rights groups and civil society organizations. These experiences continue to influence his current academic engagements.

Ricardo Fuentealba is Assistant Professor at the Instituto de Ciencias Sociales, Universidad de O'Higgins, Rancagua, Chile. He received his PhD from the University of Amsterdam in 2021 and has an interdisciplinary background in sociology, human geography and international development. His research interests concern socio-spatial inequalities, disaster risk governance, environmental conflicts and the political economy of development from critical perspectives. He has experience in collaborative and participatory work, particularly using qualitative methods in urban and rural contexts. He is a member of the collective that edits the *Journal of Disaster Studies*.

Giovanna Gioli is an independent scholar. Previously, she held academic posts in Italy, Germany and the United Kingdom, lecturing in philosophy and human geography.

Lisa Mackenzie is Senior Lecturer in the Edinburgh School of Architecture and Landscape Architecture at the University of Edinburgh and a practising landscape architect. Her research follows her commitment to understanding the potential of 'landscape' as a mediatory theory, method and practice that bridges social and environmental concerns in the climate and biodiversity crisis. Her current research explores the potential of creative practice in support of disaster risk reduction and the communication of socioecological relations in living landscapes, in forms sensitive to the complexities, inclusions and exclusions between people and nature in specific localities.

Peter McGowran is Lecturer in human geography at Queen's University Belfast. Previously, he was a research associate in disaster risk reduction and geography at King's College London and before that Lecturer in development geography at Oxford Brookes University. Peter is a human geographer interested in investigating and theorizing the relationships between disasters and development with a particular focus on majority world contexts. He has published his research in high-impact journals across geography and disaster studies. His current research is focused on multi-hazard disaster management in Kenya and East Africa. He also maintains a particular interest in and expertise on the geographies of disasters in India and South Asia.

John Hogan Morris is an assistant professor in economic geography at the University of Nottingham. He is a financial geographer with research and teaching interests in climate finance, central banking, financial risk management and equitable energy transitions. John's 2023 co-authored book, *(Mis)managing Macroprudential Expectations: How Central Banks Govern Financial and Climate Tail Risks*, explores attempts by central banks to treat climate change and low-carbon economic transitions as threats to financial stability. John serves on the steering committee of the Climate Finance for Equitable Transitions project.

Nathaniel O'Grady is Senior Lecturer in human geography and disaster at the University of Manchester, United Kingdom. His research applies a theoretical lens drawn predominantly from political and cultural geography to analyse different scenes and perspectives on disasters and their governance. His main thematic areas of focus to date have ranged from studying the techniques and rationalities that underpin government intervention in disasters, on the one hand, to exploring people's lived experiences of disasters on the other. In this vein, his work has contributed to our understanding of the politics behind the development of risk-based modes of governance, the role of digital technologies in emergency response, the formation of community resilience and, more recently, the entanglement of climate change with our everyday lives.

Ana Julia Cabrera Pacheco is a feminist geographer and decolonial scholar. Most recently, she worked as a postdoctoral research associate (2022–5) on the Ixchel Project at the University of Edinburgh. Her research expertise lies in working collaboratively with indigenous communities, particularly women, living in environments at risk of disaster and climate change in Latin America and in the multifaceted ways they respond and resist the marginalization of their ways of life. Her work has also made an impact in the areas of community engagement in disaster recovery and processes of cultural assertion through the co-production of knowledge based on land, livelihoods, plants and food.

Francisca Vergara-Pinto is an environmental anthropologist from the Andes Cordillera, Chile. She holds a master's degree in social sciences and is currently a PhD candidate in humanitarianism and conflict response at the University of Manchester, UK. Her research has explored volcanic imaginaries, disaster memory and the embodied knowledge of volcanism. She is currently focused on researching environmental humanities, geographies of affect and disaster risk, particularly in rural territories and volcanic zones. Her ethnographic and transdisciplinary doctoral research aims to deepen understanding of the affective dimensions of human-volcano relationships by examining the experiences of both local communities and volcanologists. This realm can pave the way for bridging multiple ways of knowing and for empathetically imagining safer, more vibrant futures with active volcanoes. For her, these topics are essential to understanding the more-than-human agency of geological forces and how the rhythms of volcanoes – calm, unrest, eruption and equilibrium – influence the ways in which people who are affectively engaged with them envision their everyday lives and prepare for future emergencies.

Diego Antonio Reanda Sapalú is a Guatemalan Maya Tz'utujil from Santiago Atitlán, currently a student of secondary school teaching in pedagogy and a technician in educational administration. He is a community researcher and facilitator in rural and urban areas, developing local practices and capacities and combining them with scientific knowledge. A community leader, he works to rebuild the social fabric, taking into account cultural, religious and political beliefs to create spaces for interdisciplinary and intercultural dialogue for the co-production of knowledge and tools for social well-being.

Sam Simkin is a postgraduate researcher at the University of Warwick. Her research explores the legal and institutional dimensions of climate finance, with a particular focus on gender, justice and financial governance in Small Island Developing States (SIDS). Grounded in feminist political economy, her work examines how intersecting inequalities are produced and sustained through climate finance frameworks. Sam holds an MA in international relations and a BA in politics and international studies. Professionally, she contributes to research on climate finance and global economic governance through her work with Warwick Law School's GLOBE Centre (Centre for Law, Regulation & Governance of the Global Economy) and the CLiFT project (Climate Finance for Equitable Transitions). Sam is also a member of the interdisciplinary environmental cluster based in Warwick's Department of Politics and International Studies.

Gemma Sou is Senior Lecturer in human geography at Monash University, Australia. Her main body of work draws on development geography, cultural geography and disaster research to understand how families perceive, experience and respond to disasters in their everyday lives. She also examines the politics of aid in societies affected by climate disasters as well as the theorization and production of creative research outputs. See also https://gemmasou.com/.

Elissa Waters is a political geographer and lecturer in human geography at Monash University's School of Social Sciences. Her research focuses on the governance of climate change adaptation and disasters, with particular attention to the Pacific region. She has conducted in-depth research on disaster governance in small island states, including work on sovereignty, customary authority and the politics of vulnerability in atoll contexts. In addition to her academic work, Elissa provides ongoing policy advice on climate adaptation and disaster governance in Australia and the Pacific.

Theresa Zimmermann is a research associate at the Disaster Research Unit at Freie Universität Berlin, Germany. Her research interests include the social production of risks and disasters, the agency and involvement of citizens in disaster management, and the governance of risks, disasters and resilience in the context of climate change and uncertainty. Theresa studied geography and social sciences (BSc) in Berlin and New Delhi and environmental policy and planning (MA) in Berlin. In her PhD, she analyses how disaster risk reduction is imagined, governed and negotiated in urban contexts, with a focus on flood risk governance in Mumbai, India.

Introduction

Disaster Geography's Pasts and Its Possible Futures

Nathaniel O'Grady and Gemma Sou

The pathways by which human geographical inquiry into disasters developed are strange. It is certainly the case that in some ways, the study of disasters represents a burgeoning field of research across geography. Owing in part to a changing climate, a litany of articles on disasters now appear in geography journals every year. Similarly, meetings focusing specifically on 'disaster geographies' and the 'geographies of disaster' feature regularly as the basis for workshops or sessions at geography conferences across the world. Whether appreciated or not, however, this interest in disasters is in many ways nothing new. Leafing through encyclopaedias of the discipline would reveal quite quickly that geographical research into disasters has existed for several decades. As early as the 1940s, Gilbert White was taking a geographical approach to exploring the effects of floods and their connection to plain management. Disaster research of this kind proved influential in subsequent decades in the formation of political ecology and development geographies (Watts and Bohle 1993; White et al. 2001). Disasters also continue to operate as a crucial point of focus for work that has overlapped with concerns in international relations on security, geopolitics and the politics of intervention since the 1990s and that provides the grounds for more recent work in cultural and political geography on emergency and exceptionalism (Adey and Anderson 2011; Aradau and van Munster 2011; Grove 2014; Collier and Lakoff 2015; O'Grady 2019). Across this large body of work, disasters have been analysed from a diverse set of conceptual standpoints that have structured geographical research over the years, ranging from Marxist and constructivist accounts to studies that have drawn on governmentality, more-than-human geographies and affect theory.

Despite this long lineage, a comprehensive account of the multiple connections between human geography and disasters does not yet exist. Such an omission might be explained by different reasons. One simple reason might be that the study of disasters was, for a long time, dominated by the physical sciences and applied versions of social science. Indeed, when we decided to develop this collection, we first researched what edited volumes were already available to ensure our book was a necessary addition to the literature. We soon found that the existing volumes centred on geophysical, engineering and management approaches. Another potential reason is a matter of pure abundance. As inquiry into disasters has proliferated, the relationship between geography and disasters has become increasingly fragmented and difficult to pin down. And because so much work now exists, it is perhaps too difficult and daunting a task to even attempt to address this knowledge within a unitary, coherent volume.

On a different note, there has also been much understandable reluctance to encompass all related work under the term 'disasters' specifically. A worthy question, then, is whether we even want an account that merges geography and disaster. As work on related terms like 'crisis' (Roitman 2020) and 'emergency' (Anderson 2017) suggests, the term 'disaster' has performative consequences that we should be aware of. Invoking 'disaster' serves, in other words, to shape the parameters through which we understand the wide range of events studied. Labelling an event as a disaster, for instance, might be interpreted by some to reduce its temporality to something singular and exceptional, shifting our attention away from the long-term processes that have shaped vulnerability to hazards over time. Similarly, 'disaster' can also imply that the impacts of such events take place within discrete physical boundaries when of course their impacts are often 'messy' and spill across constructed spatial limitations and scales. The term 'disaster' is often understood as a large-scale event causing widespread death, injuries and destruction. Yet others would argue that a disaster may be small scale, impacting only a small number of people or a geographical location while nevertheless leading to 'disastrous' effects for those concerned (Sou 2018). Taking this into account, scholars may have justifiably demurred from positing their research on different events under such a loaded term as 'disaster' for the immediate connotations it carries.

Another reason for potential trepidation concerns the sometimes ambivalent relations between geography and other disciplines attending to disasters.

The spatial and temporal complexity of disasters invites analysis from a multitude of theoretical approaches, and so research from various disciplines not only is necessary but should be actively celebrated. Areas in which interdisciplinary work has proven particularly useful include early warning, urban planning, hazard and risk mapping, scientific advisory processes, risk communication and institutional analyses. Nevertheless, researching disasters from a range of disciplines does come with challenges that must be acknowledged before we can elaborate upon what might be a geographically inflected approach to disasters. Interdisciplinary dialogue on disasters, which some have attempted to describe unitarily as 'disaster studies', can open up spaces where scholars intervene in debates without a full understanding of well-established ideas owing to a lack of formal training in the disciplines from which these debates emerged. In such instances, work may undermine the advancement of knowledge because it can inadvertently take us back to square one by misconstruing ideas as novel when in fact they were established many years ago. Recent work on 'disaster risk creation', the multiple temporalities of risk and the construction of risk, for instance, often fails to acknowledge that these arguments have an extensive history, exploding in the early 1990s with the work of scholars such as Ulrich Beck (2009) and Barbara Adam (2005) but dating much further back to Mary Douglas's work in the 1970s (2002). Similarly, work that claims to be agenda-setting for how we should approach disasters methodologically often replicates claims about positionality that social science accounts have been making since the late 1980s (Haraway 1989). While it is important to engage across disciplinary boundaries, then, it is equally crucial to be informed by debates that have preceded one's own work so that genuine and impactful contributions to the field can be made.

Some work undertaken in the name of interdisciplinarity can also erase the rich differences in methodological approaches, standpoints and research interests that have proven so crucial to geographical thinking about disasters. Such is the case with the reinvigoration of systems imaginaries in research on disasters of late. Systems approaches presuppose that everything within a bounded space exists in relations of interdependence with one another and that these relations will produce certain outputs that we can predict. Although complex and uncertain, only those conveniently privileged enough to understand 'the system' can objectively observe the processes by which it is made and remade, in turn prescribing managerial solutions to be applied universally regardless of the socio-cultural heterogeneity that distinguishes one place from the next. Those trained in the social sciences will recognize how such systems imaginaries, derived from cybernetics, gained prominence when integrated into the management of structurally differentiated societies in the early to mid-twentieth century (Heyck 2015; Rindzeviciute 2016). However, in recent years, systems imaginaries have been resurrected by engineering and managerial/business disciplines. Applied to interdisciplinary discussions, scholars have argued that systems imaginaries pacify difference and re-create the 'master' narratives and the truth claims that underpin them, which the social sciences have pushed against for so long (Grove and Rickards 2022). The presentation of well-established ideas as novel and the re-emergence of systems thinking are but two reasons why we should approach with caution any work that claims to 'introduce' or indeed make interventions within disaster studies or critical disaster studies, consolidating as it does a field of research that runs the risk of, purposefully or not, forgetting intellectual developments with a long history within disciplinary accounts of disasters and/or eliminating the rich diversity of knowledge and insight such disciplinary accounts possess.

Against this background, and amid a huge and diverse literature, we recognize that developing a collection that seeks to address the relationship between geography and disasters is not an easy task. But we have approached this task earnestly, first acknowledging why such a project has not been undertaken before, with the aforementioned critiques of the term 'disaster' itself in mind. We also recognize that this task requires careful reflection on both the trapdoors that feature in interdisciplinary debates on disaster and the position of geography within such debates, owing to their simultaneous importance. Considering this complicated backdrop, the book asks a number of questions. And it asks these questions not only of its authors but also to its readers. How might we characterize the relationship between human geographical thought and research on disasters? How does human geography contribute to our understanding of disasters? How can human geography position itself within interdisciplinary debates to make genuine and meaningful contributions to knowledge of disasters for all in the future?

These are questions that each of the chapters in this volume responds to in distinct ways. At the same time, there are commonalities between the chapters that, for the sake of clarity, we have encapsulated into five key themes below. Taken together, the five themes demonstrate the versatility of human geography's key concepts and subdisciplines for analysing disasters.

Core Themes

Sensing Disasters

A key theme in the edited collection concerns how geographical approaches extend our understanding of the manifold ways in which disasters are made sense of, felt and known. For some reason, recent scholarship has overlooked the etymology of 'disaster' as a word. From Latin, 'disaster' translates as the loss of one's star. This linguistic root captures not only the rupture of normative order that disruptive events like disasters engender but also how disasters complicate our ability to guide ourselves in the world. Thus, as much as the material ramifications of disaster are at the forefront of our inquiry in this book, so are the steps taken to situate such events into regimes of knowledge that underpin action in light of disasters. One way to write about sensing and knowing is as a matter of perspective and of those perspectives that the chapters in the collection adopt. Chiming with literature that has been fundamental to geography's prominence over the last few decades, one cannot understand the modes by which disasters are apprehended without of course attempting to comprehend disasters from the standpoint of those governing.

Knowing disasters is, following Foucauldian work on the topic, to inquire into the modes of problematization by which events are framed as an object of concern for governmental agencies (Collier 2009). Problematization is crucial because it allows for closer analysis of the power relations that underpin and organize governmental understandings of, and approaches to, disaster. It not only opens up consideration of the specific calculus through which disasters appear as problems and thus draws us to better comprehending the forms of knowledge that come to prevail but also opens up to critique the kinds of actions that are (or are not) sanctioned in the name of governance. Ultimately, problematization invites reflection on how disaster knowledges and interventions align with broader political-economic projects pursued by governments which simultaneously inflect our sense of the historical moment in which our inquiry takes place. In this way, we might catalogue and interrogate not only the forms of knowledge that mediate disasters from a governmental perspective but also how such knowledges become possible owing to wider processes that order our current historical moment, whether conceived as 'neo-liberal', 'mercantilist' or 'neo-colonial' and so on (see Dhungana's, Morris's, Zimmerman's contributions).

But while it leads us to broader questions of power dynamics, tracing forms of problematization is only possible with sensitivity to the local scenes from which knowledge is produced. Ironically, even as processes of problematization help us interrogate the knowledges that come to dominate, they also compel us to confront the multiplicity of relations to disaster that exist throughout the different scenes across which they unfold. Rather than only 'seeing like a state' (Scott 2020), then, sensing disasters leads us to the array of different ontologies and epistemologies that exist across communities throughout the world and how they influence people's standpoint when experiencing disasters (see Vergara-Pinto, this volume; or Waters, this volume). Existing always already beyond the putative exceptional space-time of a disaster event, these ways of knowing have developed through, and emerge from, people's daily lives. Taking this perspective allows for a more nuanced appreciation of how, when they do happen, disasters become part of individual and collective narratives that underpin people's sense of belonging to different places (Pacheco et al., this volume; Vergara-Pinto, this volume).

Geographical accounts would equally lay emphasis, however, on the fact that it is not only humans and their institutional fabulations that provide the genesis for forms of sense-making in a disaster's midst. Rather our understanding of disasters is also contingent upon our relationship with more-than-human actors. Disasters represent events in which delusions of human control over the environment are most starkly obvious. Disasters exceed human rationality and determination. Taking this into account, it should come as no surprise that the more-than-human forces entangled in, and constitutive of, disasters play an instrumental role in how disasters appear to us and how we make sense of them. While a host of elemental forces might impress themselves upon our perceptive capacities, from large volumes of water in floods to the burning flames of wildfire, our sense-making is also formed through interaction with the world that disasters leave in their wake (O'Grady 2018a). Such relations, furthermore, do not unfold in a vacuum. Instead, they can only be grasped with a keen understanding of how their configuration mirrors a much wider set of circumstances that prevail wherever we consider them.

Across these various perspectives and the actors implicated therein, the forms of knowledge mobilized in sensing disasters vary. Official discourses, ripe with statistics, taxonomic categories and reference to legislation, operate as one frame through which disasters are made intelligible and enveloped within governmental regimes. But they sit alongside knowledge that stems from people's ongoing affective, embodied encounters with disasters, whether apprehended as events that have passed, in their unfolding in a real-time durational present, or as so many risks of a future to come.

Disasters and the Materialization of Worlds

How might we conceptualize the materiality of disaster? To ask such a question is to interrogate the phenomenological nature we ascribe to disasters. Reflecting a broader canon going back in some respects to Heidegger (2010), recent scholarship has framed this inquiry through a key question: Should disasters be understood as events or as processes? At first glance, disasters seem to epitomize the notion of an event. That is to say that disasters are only recognized as such by their unsettling reverberations across a set of affairs that has proven, at least temporarily, stable. Yet geographers have cautioned that an event-centric lens risks obscuring the deeper structural conditions in which disasters are both situated and from which they arise. As such, arguments have developed that call for disasters to be considered as processes woven into the everyday and not bound by the notion of a beginning and an end. For instance, in her work on Hurricane Maria in Puerto Rico in 2017, Yarimar Bonilla (2020) unpacks the significance of a refrain common among the populace that it is colonialism that is the disaster, not the storm itself.

Rather than taking sides, this collection advances the notion that we should stop and reflect upon disasters as phenomena that hold event and process in tension with one another. Instead of one or the other, disasters take on the characteristics of both and thus exceed categorization as either. Of course, disasters need to be engaged with as events incubating with the potential to radically alter people's lives. To demur on their singularity for those in their wake would be reckless and unethical. But perhaps it would be equally so to neglect the myriad ways in which disasters reflect processes that have built up over time or instigate anew processes that integrate themselves into collective life. By embracing this duality, we resist false binaries and instead attend to disasters in their messiness as moments of abrupt devastation *and* as manifestations of deeper structural conditions.

Across the volume, the contributions tread a middle ground between process and event, creating a unique approach to the materiality of disaster. The chapters analyse with seriousness the destruction and ruin that disasters create in different places, taking stock of the consequences that they bear and their contingency on all manners of historical factors specific to a place. But at the same time, the authors elaborate on human geography's continuing and long-established emphasis on process-inflected inquiry. Accounting for the material conditions disasters both echo and actively bring about, the chapters reflect a prevailing interest too in how disasters are enveloped in processes of *materialization*. Materialization here captures disasters as inflection points: moments where latent forces, long in the making, surge to the surface. Following work that draws on Lauren Berlant's notion of situations as a particular genre of event (2011), disasters thus might be read conjuncturally, serving to sum up the historical moment to which we are contemporary. Such is the case especially when we think of how disasters such as heatwaves, hurricanes and wildfires express the wider climate crisis we are in (O'Grady 2025). At the same time, the contributions to this volume use materialization to show how disasters also catalyse new material and social formations (see Acharya and Gioli, this volume; McGowran, this volume; or Zimmerman, this volume). In this dual movement of both revelation *and* generation, disasters emerge not merely as end-points but as sites where history is made, unmade and remade.

What is constitutive of the materiality of disaster is purposefully ambiguous too. In some cases, materiality refers very explicitly to things in the world, from volcanoes and rubble to hillsides and plants. In others, materiality might be applied to trace the development of new institutional routines, forms of political agency and state-citizen relations (see Dhungana, this volume). In both forms, the approach taken in the collection situates disasters' materiality in relation to broader cultural, political and economic forces that permeate in different ways. Disasters are, in some respects, woven into some cultures as hugely destructive and integral to the belief systems that calibrate communal life (Vergara-Pinto, this volume). In other cases, disasters might be described as the latest iterations of centuries-old structural violence but also by their potential to fracture entrenched inequities and seed alternative social and political futures (see Desmaison and Fuentalba, this volume; or Pacheco et al., this volume).

The Multiple Temporalities of Disaster

Disasters are thus both disruptive events and processual phenomena, enveloped in the making and remaking of worlds and not possessing any clear beginning or end. Such a stance, of course, bears ramifications for how we think about the temporality of disaster. It might be obvious from what has preceded that we conceive of disasters as 'polyphonic' phenomena (Tsing 2015: 1). In other words, they can be attributed all manners of different rhythms and routines. For one, disasters unfold across multiple 'tenses' (Povinelli 2011), where pasts matter as much as presents and futures. It might be that it is only in a disaster's wake that we can fully assess not just the immediate damage but also how that damage is unevenly valued – revealing the biopolitical regimes that prioritize certain lives over others (Gioli and Acharya, this volume). Alternatively, perhaps the assumption that disasters are inherently chaotic and inchoate if experienced in real time warrants critique. As several chapters demonstrate (McGowran; Vergara-Pinto, this volume), such a view risks anthropocentrism and obscuring how disasters unfold as an assemblage of different more-than-human actors working in co-productive concert with one another (McGowran, this volume; Vergara-Pinto, this volume).

As one might imagine in a book about disaster, risk is a recurring theme throughout many chapters. While risk ostensibly orients us towards the future, it also functions as a temporal adhesive, binding past, present and future. So in some regards, risk ushers in and consolidates projected futures, futures that, even should they not come to unfold in the way envisioned, nevertheless bear 'real' effects by the courses of action they make possible in the here and now (O'Grady 2018b; Zimmerman, this volume). Yet risk is never a neutral calculation. As some contributors demonstrate (Morris and Simkin, this volume), risk assessments are themselves contingent on historical processes sedimented over time. The traditional calculus of risk, however, is increasingly destabilized in our climate-changed world. Where once disaster landscapes might have appeared predictable, they are now fraught with radical uncertainty. Risk is founded on a logic of probability, wherein past events are traced to understand likely futures. However, as Amoore (2013) reminds us, the future may no longer bear resemblance to the past. In this context, our imaginaries must stretch beyond probabilistic reasoning to grapple with the unprecedented, where what lies ahead may have little relation to what has come before.

But tense is only one way that the temporality of disasters might be conceived. There is also the matter of pace. Geography has in many ways become interested in the 'slow' of late. Escaping the rapid and surprising characterization of disasters with which we may have grown accustomed, these approaches emphasize how the processes underpinning, effects, and indeed the eventual occurrence of disasters themselves permeate across society with a quotidian and accretive rhythm (Anderson et al. 2020). Different sites might be bound up in slow emergencies. The meaning that people attribute to place might, in part, hinge on how they negotiate a sense of belonging amid the dangers and risks that accumulate day by day (Vergara-Pinto, this volume). Conversely, routine organizational processes can create and consolidate vulnerabilities that shape a disaster's consequences (McGowran, this volume; or Zimmerman, this volume).

The Politics of Disasters

Running across all the chapters is how geographers contribute to understandings of the politics of disaster. But what 'politics' actually refers to takes on multiple meanings across research into disasters. Politics is, in one way, deployed to identify and scrutinize the specific agendas and interests that rise to prominence in disasters and the ends they serve. Several chapters trace how specific interests manifest at different disaster stages, from immediate response to long-term recovery. This analytical lens builds on Ulrich Beck's notion of reflexive modernization, where the very pursuit of certain political and economic interests can precipitate catastrophic outcomes, as evidenced by industrial disasters such as Deepwater Horizon or Chernobyl. Equally, of course, it has become a consensus now that the aftermath of disasters needs to be studied in terms of how they are rendered as opportunities for growth and development, often at the expense of communities that had inhabited sites before.

We thus challenge the idea that "technocracy" "depoliticizes" disaster governance, arguing instead that practice is deeply imbued with politics. Interests and agendas steer, and are reflected in, the forms of intervention that governmental agencies design, curate and enact in attending to disasters. They equally undergird the arrangements, norms and conventions that consolidate governance overall (see Zimmerman, this volume; or Waters, this volume). Governmental practice, however, transcends politics as the pursuit of interests and extends also to politics as a matter of how state actors rationalize their role in collective life. Practices are indicative, in other words, of how states think about what they should, and just as importantly should not, do in times of disaster. Such rationalization in disasters is often considered to hold a mirror up to what figures as common sense about the extent to which governments should be present in collective life in general, thus underlining the political-economic philosophy steering governance.

It is important to emphasize, however, that such practices are not simply formed and mobilized at the whim of governments. Rather, they reveal the complex and evolving ties between states and citizens – ties that vary significantly across time and space. In some ways, the existence of sophisticated and elaborate forms of disaster governance might be said to be part and parcel of the promise of security that modern, liberal Westphalian states make to their people (Ophir 2007). The legitimacy of governments can hinge on their disaster governance capabilities. And failure to provide response can seriously unsettle such legitimacy too. Yet these ideals of legitimacy are absent from many contexts around the world. Many regions operate under fundamentally different political logics, where disaster governance indicates completely different state-society relations. In some instances, disaster governance can represent an intensification of the authoritarian control that states possess over citizens. In others, the absence of states in times of disaster needs to be conceived as an element of the challenges postcolonial countries face in developing adequate state capacities after independence.

In some ways, the politics of state-society relations shows the politics of disaster to be an epistemological matter as well. States seek to rationalize and legitimize their activity through particular forms of knowledge. And thus, connecting to processes of problematization discussed earlier, our inquiry is directed towards whose knowledge comes to count in disasters, why it comes to matter and the forms that it takes (see Desmaison and Fuentalba, this volume). Conversely, examining the priority afforded to some knowledges brings forth reflection upon the contestation that frequently arises concerning decisions made about how to govern disasters, whether in their incarnation as risks of the future, amidst their durative unfolding, or in their wake (see Dhungana, this volume).

Methodologies, Methods and Ethics for Understanding Disasters on the Ground in a Pluriversal Fashion

The contributions to this volume are held together by a common methodological agenda. Overall, the chapters might be said to present a pluriversal perspective on disasters. Following the work of Arturo Escobar (2018), this means that the chapters adopt an analytical stance sensitive to how experiences of disasters are intimately shaped by the diverse contexts in which they take place. The way context may be interpreted here is manifold. While the language of disaster offers a means by which to catalogue a certain type of event experienced across the world, our concern is also with a deep appreciation of treating such events in situ. To research disasters, in other words, is to engage deeply with the characteristics that render places distinct. In this view, context is not merely background but actively constitutes the significance disasters come to possess, whether that is considered in terms of collective memories, future imaginaries or the forms of sociality that they engender and organize.

Bringing this methodology to bear, the chapters in this collection showcase geographical inquiry that engages in 'on the ground' research with disaster-affected people in distinctive local contexts rather than privileging the idea that disasters can be understood from 'above' and from afar. Doing so makes for rigorous investigations that present narratives on their case with a commitment to erudite research. They take a serious approach to research that exemplifies intellectual rigour through its commitment to producing nuanced, ethically grounded narratives that counter the prevailing tendencies towards oversimplification – a problem plaguing both contemporary academia and the wider political arena. On one hand, this means that the contributions offer fine-grain perspectives on their cases. But, on the other, by exploring disasters at the so-called micro-scale, the chapters bring into focus the overlooked spaces where potential and actual transformative processes occur while simultaneously challenging top-down research that is often removed from the 'messiness' of disaster-affected contexts.

There is, of course, a particular ethical compulsion that steers this methodology. The authors in this collection show awareness of the fact that research on disasters is not an unproblematic endeavour and that ethical considerations will inevitably surface where researchers work directly with people whose everyday lives have become intertwined with disasters (see, e.g. Desmaison and Fuentalba, this volume; or Dhungana, this volume). This ethical consciousness responds to growing disciplinary concerns, following a trail of inappropriate fieldwork practices in disaster-affected contexts over many years. In this collection, all authors include a reflection on the ethics of their research, going beyond merely stating that they received human research ethics approval from their affiliated institutions. The editors understand that such reflections should be fundamental to the development of the relationship between geography and disaster and that contemporary geographical research on disasters – as demonstrated in this collection – can open opportunities for new ways to engage ethically in disaster 'zones' to guide future researchers.

Modes of Collective Life, Material Lives of Disaster and Governance amidst Disaster: Reflecting on the Individual Chapters in the Volume

While all contributions in this volume advance our understanding of the relationship between geography and disasters along these thematic lines, each does so from distinct conceptual vantage points. Some chapters might be said to foreground the modes of collective living within and beyond the state that disasters often unveil. In Ana Pacheco and her co-authors' chapter, the authors reflect on participatory fieldwork with Tz'utujil women in the Chuk Muk area of Guatemala, a key resettlement site after a deadly landslide. The chapter offers insight into how feminist geographies can help us better understand how post-disaster recovery may be strengthened through a resurgence of intergenerational agricultural practices. In so doing, it significantly extends our understanding of the myriad ways in which territories might be reinhabited after a disaster but also the emergent political possibilities such disasters can create.

Belén Desmaison and Ricardo Fuentalba's contributions also reflect on disasters' lingering aftermath but expand the analytical lens through a comparative study of urban contexts in Peru and Chile. They deploy urban political ecology as their primary theoretical apparatus to reveal how risk becomes systematically embedded within urban landscapes. They demonstrate how disaster is not merely a product of natural hazards but rather emerges from the complex interplay of historical urbanization patterns, political-economic decisions and socio-environmental transformations that accumulate over time as cities take shape.

Nimesh Dhungana elaborates on the notion of dissent as a pervasive, albeit complicated, socio-political response to such risks and the disasters that often follow. Located within a space-time produced in Nepal where the 2015 earthquake intersected with the COVID-19 pandemic, dissent is characterized here as a multifaceted phenomenon that operates simultaneously at institutional, communal and individual levels. While it undergirds and motivates the development of techniques to safeguard the most vulnerable within communities, dissent also mediates and can actively recalibrate relations between civil society and government by making space for political claims concerning the need for fairness, transparency and accountability in the delivery of public services and resources. This analysis exposes how dissent creates discursive and practical spaces for renegotiating social contracts during disasters.

These collective, civil modes of living are surrounded everywhere by the material life of disaster. And at least three chapters might be taken to make materiality their primary focus in expanding on the geographies of disaster. Giovanna Gioli and Amitangshu Acharya take up this concern by reflecting on the crucial question of the rubble produced in a disaster's aftermath. Focusing on the aftermath of the 2001 earthquake in Kachchh, Gujarat, India, the authors explore emerging discourses that distinguish between rubble and ruin, as well as the political implications of such a distinction. To divide ruin and rubble is, for the authors, simultaneously an act that renders perceivable the uneven valuation of lives affected by disaster and their situatedness within broader power dynamics that reflect not only historical forces that continue to shape a place but also the futures that might be engendered through responses to disasters in the present.

While Gioli and Acharya bring sharp focus to division, Peter McGowran's contribution might be considered as drawing our attention to the modes of coming together and entanglement that disasters often lay bare. Reflecting on research again in the South Asian subcontinent, McGowran mobilizes the concept of assemblage to take stock of, and critically inquire into, the multitude of relations between a heterogeneous ensemble of forces that orchestrate an event's unfolding. While opening up to greater appreciation of how disasters happen according to the elements that collectively constitute their occurrence as such, assemblage also engenders new arguments about a range of epistemological and ontological issues central to geographies of disaster, including the possibilities of interdisciplinary research, the sites complicit in our knowledge about disasters and a reappraisal of what we mean when we say 'there is no such thing as a natural disaster'.

Francisca Vergara-Pinto's chapter similarly grapples with the onto-epistemological questions that emerge when we ruminate on the material realities of disaster.[1] Vergara-Pinto reflects on her research across live volcano ranges in Ecuador and Chile. Drawing upon geographies of affect, she explores the different affective encounters that communities experience with volcanoes. As both actors that provide the means for living (through the landscapes they create) and simultaneously jeopardize this life (through the violent eruptions for which they are known), crucial for the author are the forms of kinship that people forge with volcanoes.

Few disasters unfold without significant government involvement, and it is around the politics of governance that the final three chapters might be said to primarily revolve. Bringing to bear economic geography into the debate, John

Morris and Sam Simkin look to the Caribbean to argue that our understanding of disasters is severely limited if we do not pay adequate attention to matters of debt and indebtedness. The authors concentrate specifically on disaster clauses that are often written into sovereign debt. These clauses can suspend a country's debt burden in the case of events of *force majeure*. Yet these provisions are more than mere financial safeguards; such clauses compel critical analysis of the calculative techniques through which loss is measured and its embeddedness in power dynamics that have not only structured Caribbean islands' relations to previous colonial regimes but also actively shaped the vulnerability to disasters in the first place.

Elissa Waters's contribution takes us to another archipelago deeply shaped by Europe's colonial legacy and how this relationship continues to reverberate through our disaster-laden present. With a changing climate, Pacific small island states are increasingly at risk from an array of devastating events that necessitate comprehensive adaptation measures. But, putting to use the vital geographical concept of scale, Waters's chapter elaborates on the difficulties, challenges and tensions that arise when universalized forms of instrumental knowledge on adaptation collide with place-specific governance realities. Her analysis lays out how generic and abstracted approaches often falter when confronted with the intricate local, place-based dynamics.

Climate change, adaptation and indeed issues of scale are at the centre, too, of Teresa Zimmerman's crucial contribution. Zimmerman's research concerns flood risk and adaptation in Mumbai. Drawing on the concept of problematization, she shows how not only are arrangements and actions pivotal to our understanding of disaster governance, but so is the framing of disasters. Such processes of problematization, however, are not necessarily as state-centric as previous literature might have us believe. Drawing on the concept of scale, Zimmerman shows instead how futures come to be imagined, and indeed acted upon, through a wide catalogue of diffuse actors who converge on a collective issue.

In the final chapter, we take stock of the volume overall. If the introduction summarizes the collection by elucidating the thematic threads that run through the contributions, the conclusion wraps up by considering them in relation to a set of wider issues that transcend the pages of this collection. It starts by ruminating on the significance that contributions bear for thinking about wider, prominent processes that play a part in constituting our present conjuncture, particularly shifting human-environment relations, the inescapability of multiple truths and epistemic relativism, continual economic crises and uneven forms of development. These explorations prove essential, we contend, for understanding how the volume charts new pathways in disaster geography. We take 'pathways' to refer to how the contributions prise open both new trajectories for future modes of inquiry *and* the kinds of challenges that might lie ahead. Our observations here will lead consequently to reflections on how the contributions impact upon geography as a discipline, on the one hand, and broader debates that some take to constitute interdisciplinary disaster studies, on the other. We will dovetail the conclusion by expanding on the importance of the contributions for developing more ethical forms of disaster research and governance.

Note

1 All interviews in this chapter have been anonymized.

References

Adam, B. (2005), *Timescapes of Modernity: The Environment and Invisible Hazards*, Routledge.

Adey, P. and B. Anderson (2011), 'Event and Anticipation: UK Civil Contingencies and the Space-Times of Decision', *Environment and Planning A* 43 (12): 2878–99.

Amoore, L. (2013), *The Politics of Possibility: Risk and Security Beyond Probability*, Duke University Press.

Anderson, B. (2017), 'Emergency Futures: Exception, Urgency, Interval, Hope', *Sociological Review* 65 (3): 463–77.

Anderson, B., K. Grove, L. Rickards and M. Kearnes (2020), 'Slow Emergencies: Temporality and the Racialized Biopolitics of Emergency Governance', *Progress in Human Geography* 44 (4): 621–39.

Aradau, C. and R. van Munster (2011), *Politics of Catastrophe: Genealogies of the Unknown*, Routledge.

Beck, U. (2009), *World at Risk*, Polity.

Berlant, L. (2011), *Cruel Optimism*, Duke University Press.

Bonilla, Y. (2020), 'The Coloniality of Disaster: Race, Empire, and the Temporal Logics of Emergency in Puerto Rico, USA', *Political Geography* 78:102181.

Collier, S. J. (2009), 'Topologies of Power: Foucault's Analysis of Political Government Beyond "Governmentality"', *Theory, Culture & Society* 26 (6): 78–108.

Collier, S. J. and A. Lakoff (2015), 'Vital Systems Security: Reflexive Biopolitics and the Government of Emergency', *Theory, Culture & Society* 32 (2): 19–51.

Douglas, M. (2002), *Risk and Blame*, Routledge.

Escobar, A. (2018), *Designs for the Pluriverse: Radical Interdependence, Autonomy, and the Making of Worlds*, Duke University Press.

Grove, K. (2014), 'Biopolitics and Adaptation: Governing Socio-ecological Contingency Through Climate Change and Disaster Studies', *Geography Compass* 8 (3): 198–210.

Grove, K. and L. Rickards (2022), 'Contextualizing Narratives of Geography's Past, Present, and Future: Synthesis, Difference, and Cybernetic Control', *Environment and Planning F* 1 (1): 26–40.

Haraway, D. (1989), 'Situated Knowledges: The Science Question in Feminism and the Privilege of Partial Perspectives', *Feminist Studies* 14 (3): 575–99.

Heidegger, M. (2010), *Being and Event*, State University of New York Press.

Heyck, H. (2015), *Age of System: Understanding the Development of Modern Social Science*, Johns Hopkins University Press.

O'Grady, N. (2018a), 'Communication and the Elemental: Capacities, Force and Excess in Emergency Information Sharing', *Environment and Planning D: Society and Space* 37 (1): 158–76.

O'Grady, N. (2018b), *Governing Future Emergencies: Lived Relations to Risk in the UK Fire and Rescue Service*, Springer.

O'Grady, N. (2019), 'Designing Affect into Security: Shared Situational Awareness Protocols and the Habit of Emergency Response', *Cultural Geographies* 26 (4): 455–70.

O'Grady, N. (2025), '(Re)framing the Politics of Climate Change: Resilience, Reparative Critique and Affective Life in Situations of Extreme Heat', *Critical Studies on Security* 13 (1): 70–86.

Ophir, A. (2007), 'The Two-State Solution: Providence and Catastrophe', *Theoretical Inquiries in Law* 8 (1): 117–60.

Povinelli, E. A. (2011), *Economies of Abandonment: Social Belonging and Endurance in Late Liberalism*, vol. 48, Duke University Press.

Rindzeviciute, E. (2016), *The Power of Systems: How Policy Sciences Opened Up the Cold War World*, Cornell University Press.

Roitman, J. (2020), *Anti-Crisis*, Duke University Press.

Scott, J. C. (2020), *Seeing Like a State: How Certain Schemes to Improve the Human Condition Have Failed*, Yale University Press.

Sou, G. (2018), 'Mainstreaming Risk Reduction into Self-Build Housing: The Negligible Role of Perceptions', *Climate and Development* 10 (6): 526–37.

Tsing, A. (2015), *The Mushroom at the End of the World: On the Possibility of Life in Capitalist Ruins*, Princeton University Press.

Watts, M. J. and H. G. Bohle (1993), 'The Space of Vulnerability: The Causal Structure of Hunger and Famine', *Progress in Human Geography* 17 (1): 43–67.

White, G. F., R. W. Kates and I. Burton (2001), 'Knowing Better and Losing Even More: The Use of Knowledge in Hazards Management', *Global Environmental Change Part B: Environmental Hazards* 3 (3): 81–92.

1

Dissenting in Disasters

Lessons for the Disaster–Democracy Interface from Nepal's Dual Disasters

Nimesh Dhungana

How does the politics of dissent manifest in the wake of disasters? Dissent as a concept remains central to studying democracy, but as of yet, it has been minimally deployed by geographers to examine democratic citizenship more broadly (Staeheli 2010) and to consider the nature of political contestation in the everyday context of disasters and its implications for the democratic governance of disasters. Against this background, this chapter bridges debates on democracy and disasters through the lens of dissent. It does so by exploring Nepal's dual disasters: the 2015 earthquakes and the COVID-19 pandemic.

Scholars have long considered disasters as collective stress situations that call for societal consensus and cooperation to alleviate widespread suffering (Barton 1969). Contemporary examples of societal consensus and cooperation include the internationalization of disaster risk reduction (DRR) and recovery, in which states and international actors make explicit or implicit reference to the normative values of state–societal cooperation, encapsulated in policy rhetoric such as 'development through disaster' or 'building back better'. The aftermath of disasters is also marked by authorities calling for civic assistance and compliance. This was evident during COVID-19, with politicians using language such as 'we are all in this together' to evoke a sense of collective disruption and the need for compliant societal behaviour (Prime Minister's Statement on Coronavirus (COVID-19) 2020). Prime Minister Narendra Modi of India called for a nationwide lights-off vigil as a symbol of national unity to counter what he termed 'negative sentiments' (BBC 2020). The Nepalese prime minister Sushil Koirala made similar calls for unity following the 2015 Nepal earthquakes, stressing the need for 'partnership' and 'voluntarism' amidst the national tragedy (*The Kathmandu Post* 2015).

Yet the cooperative societal behaviour that authorities often expect from citizens does not always transpire during disasters. Indeed, disasters are also a fertile ground for political struggles over how to govern the disaster, by whom and to what ends, making response and recovery efforts inherently conflictual (Stallings 1988). Disasters are marked by the politics of fault-finding and blame-making, with authorities facing public outrage for failing to prevent and mitigate the impacts of disasters (Drabeck and Quarantelli 1967). Tensions between state-driven narratives of collaboration and compliance and community-driven initiatives of compassion and care are characteristic of disaster response (Xu 2017; Matthewman and Uekusa 2021). Using a mix of theoretical insights ranging from social capital and participatory democracy to social movements, scholars have examined the unfolding of creative forms of citizen-driven initiatives that are centred on promoting the values of transparency, accountability and justice in the wake of disasters (Luft 2009; Xu 2017; Cretney 2019).

Such contentious politics in the aftermath of disasters have also been the subject of geographical scholarship, offering new insights into the nature of state-societal tension and cooperation that tend to shape longer-term recovery initiatives (Pelling and Dill 2010; Cretney 2019; Siddiqi 2019; O'Grady and Shaw 2023). For Pelling and Dill (2010: 2), disasters tend to spark 'provocative social processes at multiple social levels', often posing serious challenges to the dominant form of state-led disaster governance. Following the 2010 Christchurch earthquake, Cretney (2019) highlights the emergence of the 'politics of possibility', evidenced in community-driven disaster recovery to counter the risk of depoliticization and administrative disregard for the lived realities of disaster-affected communities. Luft's (2009) longitudinal ethnographic study of recovery efforts following Hurricane Katrina also provides a vivid account of how post-disaster social movements seek to reverse the state's attempt to frame disaster as an 'exceptional event' by spotlighting its linkages with deep-seated racial and structural inequalities. According to Solnit (2009), the political possibility of disasters involves the renewal of civil society, an indispensable part of the politics of hope, despite the collective disruption wrought by disasters. Likewise, Nicole Curato's (2019: 54) ethnographic inquiry into the 2013 Typhoon Haiyan in the Philippines highlighted the innovative role of 'surrogate publics', spanning digital innovators, community organizers and activists who tried to reverse the prevailing power inequalities between state authorities and local communities. Beyond their engagement in alleviating the immediate suffering of disaster-affected communities, these empirical insights highlight the role of innovative forms of citizen engagement in countering the possibilities of power misuse or abuse in the guise of a 'state of exception'.

Dissenting in Disasters: Possibilities and Challenges

A key element of such political organizing in the aftermath of a disaster is public dissent. The concept of dissent, often overlooked in geographical discussions about the emergence of public spheres, spans a range of cooperative and confrontational modes of democratic organizing (Staeheli 2010). Dissent, thus, serves as an interesting lens to explore the intricate relationship between democracy and disaster.

According to the Cambridge Dictionary, dissent refers to the act of opposition about a particular subject, especially involving officials or authorities. Although dissent as an oppositional politics is often traced to the values of the Enlightenment and liberalism stemming from the West, the politics of dissent is hardly limited to Western democracies (Sen 2006; Graeber 2007). Non-Western societies have long favoured and pursued structures of governance that give primacy to dissenting public spheres. Civil disobedience (Satyagraha, or pursuit of truth), championed by Gandhi (n.d.) in India, is one such example, illustrating the power of non-violent and passive forms of civic resistance to challenge and dismantle colonial infrastructures of governance.

Despite a long and vibrant history of contentious forms of grassroots political mobilization in the Global South, mainstream studies of politics have given less attention to such subaltern approaches to securing freedom and justice or what Escobar (2020: 5) calls pluriversal politics for alternative societal change. Escobar's argument aligns with what Kunreuther (2018: 24) terms the 'sounds of democracy', bringing to the fore the South Asian concept of 'awaj' (voice), which she defines as 'a collective expression of political consciousness'. Awaj distinguishes itself from conventional voting, which is typically regarded as an episodic manifestation of individual political preferences within representative democracies. In contrast, awaj involves a more sustained and collective approach to democratic practice aimed at critically scrutinizing and questioning the actions of power holders, including, but not limited to, elected representatives.

While the nature of awaj, or voice-based politics in disaster settings, remains underexamined and serves as a significant empirical entry point for this chapter, some theoretical insights can be gleaned from the existing literature on disaster politics. As Olson (2000: 154) argues, 'in any disaster, government officials are confronted with the need to not only manage the situation but also explain it', indicating how power holders become vulnerable to the pressures of the critical citizenry in the face of a crisis. In a similar vein, Birkland (1998) has characterized disasters as 'focused events', offering a space for disadvantaged groups to reclaim their fundamental rights to be protected against disasters. More generally, such political claim-making in the wake of disasters reorients its focus on 'democratic values and principles, as well as in the mechanisms designed to safeguard them' (Svedin 2012: 14). In other words, the act of dissent is focused not just on demanding the state's protection from disasters but also on defending and promoting the values of democratic citizenship itself while casting a critical spotlight on the capacity and credibility of the state to protect its citizens from harm.

The normative and instrumental value of 'awaj' (voice) as an act of public dissent can be further traced to capability-based development scholarship. Amartya Sen's seminal work on democracy and response to famine (Drèze and Sen 1989) brought the disaster-mitigating value of voice to the forefront of academic and policy discussions. According to Drèze and Sen (1989), 'public action', spanning citizen-led advocacy and alert campaigns, acts as a major safeguard against disasters. India's success in averting large-scale famine is attributed to the democratic opening and the environment of public scrutiny that followed its independence in 1947. In their later work, Drèze and Sen (2002) have redefined the concept of voice in terms of twofold action: (a) self-assertion by marginalized communities and (b) solidarity among progressive social actors (i.e. social movements and political activism). Both forms of voice – self-assertion and solidarity – are considered to have intrinsic, instrumental and constructive value, especially in translating the abstract notions of rights into concrete participatory activism to overcome the chronic problems within the state-driven public sector governance (e.g. corruption, low quality of services, chronic unavailability and absenteeism of public officials).

Notwithstanding these potentials, critical literature on voice has also considered the barriers facing voice-based politics in improving human well-being. According to critical media scholar Nick Couldry (2010: 10), voice should be seen not only as an individual expression of giving an account of oneself but also as a form of collective action, indicating the existence of entrenched structural inequalities that hinder the exercise of voice in the first place – a concept he refers to as 'voice-denying rationality'. Through extensive examination of the unfolding of local public hearings in India, Sanyal and

Rao (2018) further theorized the grassroots expressions of voice as a form of 'oral democracy', a collective struggle to force local authorities to tackle everyday forms of governance gaps (e.g. funding misallocation, infrastructural mismanagement, drinking water shortages, quality of education).

Dissenting by Whom, for What and to What Ends?

Despite these vibrant debates, geographical literature has been slow to theorize and examine what it means for disaster-affected communities to dissent in the wake of disasters. Disasters are considered 'totalising events', triggering consistency and inconsistency, coherence and contradiction, cooperation and conflict, hegemony and resistance (Oliver-Smith 1999). The resurgence of political possibilities and challenges following disasters demands a critical assessment of the normative assumptions about dissent as a form of vocal expression of opposition and a deeper understanding of the systemic barriers against voice-based political action. Understanding the deep-seated obstacles to voice-based politics is key, as governments are known to consolidate their power in the wake of major disasters, constraining the spaces for inclusive and citizen-based disaster recovery (Hayward and Johnson 2022). As Erica Caple James (2010: 223) argues through her ethnographic account of post-Haiti earthquake reconstruction, while disaster-affected citizens may have the agency to speak, their voice may not bring about improvements in their well-being or 'security' when rights and voice are merely used to uphold and expand the bureaucratic reach of disaster reconstruction. James's findings challenge the normative foundation of the value of voice in enabling ordinary citizens to address what Sen has termed 'freedom' and 'security' from manifest forms of material and political deprivation.

Critical analysis of dissent is also warranted in the context of the growing adoption and internationalization of the notions of voice and accountability in DRR. In recent years, both government and non-governmental organizations (NGOs) have introduced various mechanisms to capture and cultivate the voices of disaster-prone communities to improve humanitarian aid. Such mechanisms range from complaint-handling or redressal mechanisms to community hotlines. However, emerging literature shows that such formal modes of complaint-handling mechanisms consider voices as feedback, treating disaster-affected communities as mere customers of humanitarian aid rather than citizens with rights and claims (Madianou et al. 2016; Dhungana 2020a). Such logic of project- or expert-driven approaches to public voice means that critical scholarly attention is necessary to understand how official interpretations of voice-based DRR compare and contrast with informal and everyday expressions of civic voice.

Research Question and Methodology

The rest of the chapter utilizes the aforementioned theoretical insights on dissent to illustrate the broader landscape of dissenting citizenship following disasters. The aim here is to draw on and expand the existing geographical insights on state-societal politics under challenging contexts, further illustrating how the dissenting public poses newer forms of challenge to state-driven disaster governance. The chapter utilizes the 'extended case method' (Burawoy 1998) and focuses on two distinct yet comparable disasters in Nepal: the 2015 Nepal earthquakes and COVID-19. According to Burawoy, an extended case study involves a reflexive method in which the researcher engages in a sustained dialogue with the research participants, combined with ethnographic accounts of the context in which everyday societal activities unfold. Such dialogue-oriented research calls for longer-term engagement with the ontological context, providing a basis to make broader theoretical claims about the political dynamics of disasters and their aftermath. Detailed findings from the interview and ethnographic research on these two disasters have been published elsewhere (Dhungana 2020a, 2021) or are in the process of being published. In this chapter, I aim to reorient our attention to the concept of dissenting in disasters, informed by two interrelated empirical questions: How do the politics of dissent manifest in the aftermath of disasters? What possibilities do such politics hold in advancing a just and democratic governance of disasters?

In line with the extended case method, the analysis draws on ethnographic inquiries into the unfolding of various forms of collective action and participatory activities in the wake of the 2015 earthquakes and COVID-19, together with interviews with disaster-affected communities, civil society actors, disaster management professionals and policy advocates based in Nepal. The chapter is also informed by my longitudinal and collaborative engagements with youth-led civil society organizations in Nepal, namely, Accountability Lab, which specializes in grassroots initiatives to strengthen social accountability in Nepal's development and disaster sectors. Much of the empirical insight is drawn from a range of grey literature (reports, press releases, petitions) produced by youth activists in Nepal. Critical engagement with such 'activist perspectives', as has been done by disaster sociologists (Luft 2009; Xu 2017), is consistent with my interest in moving beyond scientific or expert knowledge to account for the resurgence of unconventional forms of activist- and practitioner-driven knowledges made possible by disasters.

Nepal's Dual Disasters: 2015 Nepal Earthquakes and COVID-19

In April and May 2015, an earthquake with a magnitude of 7.8 and its aftershocks proved to be a major humanitarian disaster for Nepal. It claimed over 9,000 lives and caused major disruption to Nepal's development aspirations, from which the country is still recovering. In late 2019, Nepal faced another humanitarian crisis with the outbreak of COVID-19. Although the 2015 earthquake and the COVID-19 outbreak differed in their origin, scale and scope of societal impacts, they shared some common characteristics, making an extended case approach a suitable methodological and analytical choice for this research. First, both disasters disproportionately impacted disadvantaged communities such as the urban poor and migrant workers. This, in turn, calls for a longitudinal inquiry into the emergence and evolution of dissenting publics and their role in safeguarding the rights and welfare of marginalized communities. Second, both disasters served as an unveiling force, bringing into public discourse the governance dysfunctions and deficits that preceded and worsened the unequal impacts of these dual disasters. Third, in response to the governance dysfunctions, both disasters saw the emergence of citizen-led politics of voice and participation, providing a suitable empirical context to capture the characteristics and consequences of such politics and its implications for policy and practice.

The following section summarizes how the politics of dissent took shape following Nepal's dual disasters, building on established and emerging geographical insights into the politics of state-societal relations in the wake of disasters. The chapter concludes by discussing the implications of these findings for the study and practice of the disaster–democracy interface.

Challenging and Decentring Centralized Disaster Governance

In Nepal's dual disasters, the politics of dissent manifested in disputes over official and expert-led emergency responses. Both disasters saw the government of Nepal using its executive powers to declare a state of emergency, enabling the Home Ministry to take a decisive role in implementing rescue and relief measures. Under the Home Ministry's direction, the 2015 Nepal earthquakes saw the deployment of security forces to rescue people, clear debris and distribute relief materials. Similar forms of state-led response also characterized the COVID-19 response, with security forces directing their attention to imposing strict lockdowns and border closures that, in turn, posed a serious threat to the rights and dignity of international migrant workers.

Such centralized and state-led emergency response was soon challenged by emergent forms of citizen action, deeming the former grossly inadequate in alleviating the immediate humanitarian suffering. Following the 2015 Nepal earthquakes, spontaneous groups of first responders and volunteers organized relief and rescue efforts. In my interviews with youth volunteers, they spoke passionately of clearing the debris, raising funds and mobilizing food and temporary shelters to communities, while the government scrambled to mount its relief operations. In contrast to the government-led assessment of victimhood, that is, the formal identification of victims and eligibility determination for government assistance, activists pursued an informal assessment of victimhood and deservedness of aid. Such efforts, as my ethnographic accounts from Sankhu, a peri-urban community on the outskirts of Kathmandu Valley, showed, sought to challenge and redefine the government's eligibility criteria for the reconstruction of damaged residential houses and what the government termed 'owner-led reconstruction' (Dhungana 2021). By organizing local meetings that often saw the participation of reconstruction experts and officials representing the state-mandated National Reconstruction Authority (NRA), civil society workers demanded a redefinition of the official eligibility and questioned the rhetoric of 'owner-led reconstruction'.

COVID-19 saw similar acts of grassroots organization of mutual aid groups despite the strict lockdowns imposed by the government. My collaborative work with a Nepal-based youth-led organization, Accountability Lab, has documented local activists providing transport support to domestic migrants who were stranded following the abrupt imposition of strict lockdowns and who also came under close surveillance by security forces. Civic renewal of this kind transcended local and global boundaries, aligning with what Darling (2009) refers to as 'relational politics', an outward-looking and place-defying political and moral endeavour mobilized in the interests of those deemed excluded from the state's relief system. As one example, Nepalese international migrant groups faced disproportionate impacts of the pandemic in the form of loss of employment and fear of deportation from their destination countries, prompting migrant activists and rights groups to organize online relief campaigns that defied the situation of, as one migrant recalled, 'absent government'. As the formal expertise of disaster response became overwhelmed, the Nepalese diaspora came together to raise funds to provide medical supplies such as medical oxygen and personal protective equipment.

Back in Nepal, youth groups worked to tackle black marketing and exposed malpractices related to shortages of essential commodities. In particular, the second wave of COVID-19 saw a major scarcity of essential medical commodities such as medical oxygen. Against such a dire situation of scarcity, Covid Connect Nepal, a youth-driven civil society group, brought together digital volunteers to connect COVID-19-affected individuals with health professionals and plasma donors, while identifying and supplying oxygen cylinders to desperate families.

As noted by O'Grady et al. (2022), the emergence of such communities during disasters characterizes multiple and even contrasting functionalities when state-led emergency governance is under stress. Disaster-affected communities have the potential to simultaneously expose the inconsistencies and deficiencies in state-driven emergency governance while mediating the interactions between the affected communities and governmental authorities.

Questioning Official Narratives and Countering the Risk of Power Abuse

In addition to posing a major challenge to the state-led centralized emergency response, both disasters sparked a web of activism aimed at questioning and reversing the government's handling of the crisis. One such effort involved public scrutiny of the 'One Window' policy introduced by the government of Nepal in the immediate aftermath of the Nepal earthquakes. Civil society actors questioned not only the intent of this policy to bring widely dispersed relief efforts within government jurisdiction but also the opaque process through which it was introduced. The opposition involved local activists and the Nepali diaspora putting together a memorandum signed by thirty-six Nepal-based civil society organizations/NGOs demanding the government 'not to stick on its stance on one door system based on the nature of rescue and needs of earthquake-affected community', while also cautioning the government against 'curtailing people's constitutionally guaranteed rights' (Civil Society Organisations and NGO Federation of Nepal 2015). Another initiative that was the subject of my investigation involved a network of 'social auditors' who quickly came together after the 2015 earthquake to track the flow of government relief operations. But it was not limited to that. The social auditing involved exposing what the activists termed 'State amnesia', referring to the state's tendency to avoid prioritizing DRR despite Nepal's vulnerability to disasters (Raj and Gautam 2015).

Interview data further showed civil society groups using social media platforms to name and shame public officials for their failure to expedite relief operations. Defying the official narratives of 'victimhood', the earthquake-affected communities rallied around their constitutionally guaranteed right to protest, with some groups organizing sit-ins and demonstrations in Kathmandu and other places against the opaque and exclusionary form of government relief operations.

Like the Nepal earthquakes, the COVID-19 pandemic stirred political opposition transcending local and global boundaries. The border closure and ensuing hardship faced by returnee migrant workers served as an 'attention-grabbing incident' (Birkland 1998), with both mainstream and social media making visible the acute suffering and deprivation facing the stranded migrants. This, in turn, invigorated civil society and diaspora activism, questioning the official justification behind a harsh and one-size-fits-all lockdown. In some instances, international labour migrants organized demonstrations outside the Nepalese Embassy, demanding that the government fulfil their fundamental right to return to the country (Onlinekhabar 2020). In June 2020, the pandemic-induced political dissent reached new heights. Nepali youth took to the streets under the slogan 'Enough is enough', staging socially distanced protests against the government's slow response to the pandemic's deepening socio-economic toll (Aryal 2021). The campaign successfully compelled the government to reach a twelve-point agreement pledging to improve testing and treatment services while also ensuring transparency and accountability in the government's response.

During both the 2015 Nepal earthquakes and COVID-19, online platforms such as Twitter (now X) and change.org were widely leveraged to amplify people's voices about alleged misuse and corruption in delivering aid. One such act involved civil society rallying around the government's alleged abuse of power and corruption in procuring medical equipment from China. Challenging the government also involved civil society-led efforts to dispel misinformation and myths circulating in the heightened uncertainty following the pandemic. Attempts were made to tackle 'misinformation that circulated like wildfire', as one interviewee put it. For instance, Shramik Sanjal, an online community of international migrants, took to their Facebook account to dispel rumours and misinformation that they considered exposed fellow migrants to health hazards. These myriad forms of digital organizing, some spearheaded by those considered 'vulnerable' in the official narratives, represent a wider struggle to reestablish people's 'right to know' and 'be heard' in the face of a major crisis. The political activism around the right to information was also evident in youth groups organizing online monitoring, for instance, to track the supply of medical equipment such as oxygen cylinders. Such repertoires of offline and online activism follow insights from Hayward and Johnson's (2022) study of democratic renewal following the Christchurch earthquake, countering the risk of authoritarian tendencies taking root in the name of difficult and exceptional circumstances.

Activating State-Centric Structures of Oversight and Accountability

Nepal's dual disasters were not merely limited to the bottom-up and citizen-driven mobilization of societal discontent. They also saw public pressure to activate some key state institutions to regulate the conduct of the government.

At the national level, the aftermath of the 2015 earthquakes saw the active involvement of the National Human Rights Commission (NHRC), a constitutionally mandated oversight body. Amid growing concerns over the possibility of the government sidelining people's rights, the NHRC undertook field-level monitoring of the government's relief operations and called on the government to ensure rights-based disaster recovery. During COVID-19, the NHRC again sought to further the politics of the right to information in favour of affected communities, mobilizing its nationwide network of rights activists to monitor the situation of migrants and calling on the government to ensure that returnee migrants were provided with adequate health check-ups, food and shelter, and ensured 'dignified repatriation to their homes' (National Human Rights Commission of Nepal 2020: 2). Likewise, during COVID-19, rights activists organized a legal campaign to challenge the government's decision to close international borders and the only international airport, which interfered with migrant workers' right to return home. The decision was deliberated in the Supreme Court, with the court urging the government to ensure the safe and dignified return of the migrants while also guaranteeing adequate testing and treatment provisions. Similar tactics of building progressive state–societal alliances were also mentioned in the interviews, with the migrant activists pressurizing the parliamentarians to bring migrants' rights and well-being to the forefront of parliamentary deliberation.

My ethnographic inquiry into the unfolding of local-level public hearings demonstrated that the complex and opaque structures of centralized reconstruction efforts meant that disaster-affected communities remained uninformed about who was responsible for what and how to claim assistance from the government. At the local level, youth activists organized public hearings, offering the disaster-affected communities awaiting government assistance a way to mount pressure on the authorities. At the same time, such participatory spaces also represented the efforts of local activists to reduce the mistrust facing local authorities, especially as their political capacity to steer the reconstruction became highly questionable amidst the growing centralization of reconstruction efforts under the command of the NRA, a central body formed to oversee the recovery efforts. Public hearings were also organized at the local level during COVID-19, with youth activists trying to develop working relationships with local officials to institutionalize social protection measures for returnee labour migrants (Figure 1.1).

Figure 1.1 A public hearing in Sankhu (Sankhrapur Municipality) was a site of state–societal encounter; the local officials came face to face with the affected communities to respond to their concerns and complaints surrounding delayed reconstruction.
Source: Author.

The above insights indicate a distinct form of state–societal politics emerging after Nepal's dual disasters. While the previous themes indicate that citizen-driven politics of dissent emerged, at least in part, from a lack of trust in the functioning of state-led executive structures – 'the absent government' as one interviewee put it – the theme of 'activating state structures' reveals that citizens have hardly abandoned their democratic aspirations for the state to advance an inclusive and accountable approach to disaster recovery.

Cost of and Challenges to Dissenting in Disasters

In investigating how dissenting publics emerged and evolved in the everyday context of Nepal's dual disasters, the study also revealed different ways in which power-holders were adept at advancing and justifying counter-narratives of centralized governance of disaster recovery and, in the process, discrediting or undermining opposing voices. Put simply, the risk of erosion of dissenting citizenship represents another reality of the disaster's aftermath that warrants critical inquiry. Ethnographic observations of the local meetings and public hearings revealed how reconstruction officials refuted demands for information and transparency from local activists, labelling them, for example, as 'uncooperative' and 'politically motivated', indicating a lack of political will on the part of the state authorities to engage with, let alone cultivate, contentious voices. Likewise, interviews with earthquake-affected communities demonstrated how local acts of opposition ran the risk of authorities making it harder for dissenting individuals to claim their recovery entitlements such as housing grants.

Some of my interviewees expressed anxiety over the possibility of having to wait longer to receive their housing assistance if deemed too vocal or confrontational by the officials. Similar insights were also expressed during COVID-19 by returnee migrant workers, with some choosing to avoid confronting local authorities due to the fear of government backlash in the form of unnecessary hassle in getting their paperwork authorized for future overseas labour migration. Confronting the authorities, as per these insights, is far from being mere act of vocal opposition but a negotiated and delicate political act, especially under unequal state–societal relations. Fearing backlash from the authorities, disadvantaged communities often opt to temporarily retreat from lodging their vocal disapproval. But this does not mean they have chosen to succumb to the government-imposed recovery and reconstruction approaches. Nor does this mean abandoning their expectations of the state, as discussed above regarding the civic struggles to activate state structures. Instead, this represents a complex and intelligent act of citizenship under an uncertain and fluid political climate of disaster recovery.

Conclusion

That disasters are not just moments that give rise to altruistic support networks but also politically contentious spaces is increasingly acknowledged. However, compared to other social science disciplines such as political science and sociology, geographical scholarship has been slow to explore the nature of political dissent in the everyday context of disasters and its implications for democratic governance of disasters. Drawing on my longitudinal and extended case analysis of the 2015 Nepal earthquakes and the COVID-19 pandemic (Dhungana 2020b), this chapter seeks to address this gap.

Despite the widespread social disruption, Nepal's dual disasters were marked by a plurality of dissenting citizens who contested the official narratives of victimhood and vulnerability. Disaster-affected communities, working in solidarity with civil society groups, asserted their rights as citizens and, in the process, advanced informal and heterogeneous forms of solidarity movements. Such acts of decentred civic engagements defy the simplistic victim–survivor framework that mainstream disaster studies tend to deploy. Rather, as recent geographical scholarship has shown, disaster-affected communities tend to contest official narratives of 'building back better', striving to reclaim recovery models that address historical vulnerabilities and socio-economic marginalization (Cretney 2019; Rhiney 2020). Nepal's dual disasters involved similar politics of claim-making, evoking a sense of commitment and care towards the rights and well-being of communities in crisis. The political response to reclaim the nature of response and recovery is reminiscent of what Appadurai (2004) calls 'future-oriented politics', a deep and intelligent form of citizenship under difficult and uncertain conditions. The chapter argues that dissenting in disasters involves nebulous modes of societal aspiration for just and democratic recovery and reconstruction from disasters.

Yet such future-oriented politics is not independent of the retrospective assessment of the ongoing contextual conditions and governance dysfunctions that produce or exacerbate disasters in the first place. Nepal is not new to political upheavals and associated state-societal tension. In particular, the past two political movements (1990 and 2006) are marked by lofty promises from the authorities to improve public participation and devolve governing authority to local bodies, including the governance of DRR. However, such promises are not matched by a shift in power from the centre to the local, nor by investment in the public service delivery capacity of local bodies (Acharya and Zafarullah 2020). Despite the regular threats of disasters, the state-led disaster governance apparatus suffers from a chronic lack of political will and regulatory power to bolster participatory and community-based risk mitigation and recovery models (Karki and Salike 2020). Although the recent political changes in Nepal have opened up newer spaces for citizen engagement, citizens' political agency to bring radical shifts in the decisions and actions of the government remains questionable. In such a challenging context, expressing discontent over the state-led disaster response is not a linear or uninterrupted political action. I argue that collective acts of disapproval are susceptible to falling prey to bureaucratic stubbornness and even backlash. Mobilization of dissent in disasters, thus, also means maintaining and sustaining the 'navigational capacity' of the citizenry, capable of tracing the roots of deep-seated power hierarchies (Appadurai 2004).

In disaster contexts, as the chapter shows, mobilizing dissent may mean civil society actors forging tactical alliances with local officials. Such state–societal alliances may also span efforts to activate the enforceability power (or teeth) of the state-mandated judicial and oversight structures such as the NHRC or the Supreme Court. Building on Ayesha Siddiqi's (2019) ethnographic critique of the commonly held understanding of 'the absent state' in the aftermath of the 2010 floods in Pakistan, Nepal's experience indicates the constructive and progressive role that state oversight and regulatory bodies can play in promoting just and accountable disaster recovery. Specifically, the chapter calls for further research into the hitherto limitedly understood aspects of the inner workings of the state–societal alliances towards a meaningful DRR.

In sum, dissenting in disasters is hardly a distraction from the immediate demands to alleviate human suffering. Rather, dissenting as a form of 'awaj' (defined above) relates to what philosopher and human development scholar Martha Nussbaum (2018) terms the politics of hope amidst an environment of insecurity and despair. To be able to dissent in disasters is also about creating a possibility for reforms, an alternative politics of public imagination that may diverge from or defy the dominant or official narratives of 'building back better'. The chapter also has implications for the methodological dimensions of disaster research. It underscores the potential of extended and longitudinal case study design for the study of disasters, paying closer attention to comparative ethnographic accounts of the lived realities of disaster-affected communities, together with plural and alternative knowledges developed and advanced by disaster activists. Activist knowledges have the potential to reveal the complex yet promising interplay between 'micro' (e.g. youth-led demonstrations) and 'macro' (e.g. diaspora activism) levels of democratic invigoration in the face of a major disruptive phenomenon. Disaster scholars should pay further attention to the longitudinal and multi-scalar analysis of disaster aftermath to bring to public attention the less visible or hidden acts of civic opposition within and across different phases of disaster governance (preparedness, response, recovery and rebuilding).

References

Acharya, K. K. and H. Zafarullah (2020), 'Institutionalising Federalism in Nepal: Operationalising Obstacles, Procrastinated Progress', *Public Administration and Policy* 23 (2): 125–39. https://doi.org/10.1108/PAP-03-2020-0013.

Appadurai, A. (2004), 'The Capacity to Aspire: Culture and the Terms of Recognition', in V. Rao and M. Walton (eds), *Culture and Public Action*, 59–84, Stanford University Press.

Aryal, T. (2021), 'Will the Nepali Revolution Be Tweeted?', The Record, 12 March. Available at: https://www.recordnepal.com/will-the-nepali-revolution-be-tweeted (accessed 7 November 2025).

Barton, A. H. (1969), *Communities in Disaster: A Sociological Analysis of Collective Stress Situations*, Doubleday.

BBC (2020), 'Coronavirus: India Holds Lights-Off Vigil as Modi Calls for Unity', BBC News, 5 April. Available at: https://www.bbc.com/news/world-asia-india-52177506 (accessed 7 November 2025).

Birkland, T. A. (1998), 'Focusing Events, Mobilization, and Agenda Setting', *Journal of Public Policy* 18 (1): 53–74. https://doi.org/10.1017/S0143814X98000038.

Burawoy, M. (1998), 'The Extended Case Method', *Sociological Theory* 16 (1): 4–33. https://doi.org/10.1111/0735-2751.00040.

Civil Society Organisations and NGO Federation, Nepal (2015), 'Joint Statement from Civil Society Organizations', 30 April. Available at: https://imadr.org/wordpress/wp-content/uploads/2015/05/Nepal_Joint-Statement-from-Civil-Society-Organizations_30April2015.pdf (accessed 7 November 2025).

Couldry, N. (2010), *Why Voice Matters: Culture and Politics After Neoliberalism*, Sage.

Cretney, R. (2019), '"An Opportunity to Hope and Dream": Disaster Politics and the Emergence of Possibility Through Community-Led Recovery', *Antipode* 51 (2): 497–516. https://doi.org/10.1111/anti.12431.

Curato, N. (2019), *Democracy in a Time of Misery: From Spectacular Tragedies to Deliberative Action*, Oxford University Press.

Darling, J. (2009), 'Thinking Beyond Place: The Responsibilities of a Relational Spatial Politics', *Geography Compass* 3 (5): 1938–54. https://doi.org/10.1111/j.1749-8198.2009.00262.x.

Dhungana, N. (2020a), 'Doing Civil Society-Driven Social Accountability in a Disaster Context: Evidence from Post-Earthquake Nepal', *Politics and Governance* 8 (4): 395–406. https://doi.org/10.17645/pag.v8i4.3154.

Dhungana, N. (2020b), 'Human Dignity and Cross-Border Migrants in the Era of the COVID-19 Pandemic', *World Development* 136:105174. https://doi.org/10.1016/j.worlddev.2020.105174.

Dhungana, N. (2021), 'The Politics of Participatory Disaster Governance in Nepal's Post-Earthquake Reconstruction', in M. Liechty, M. Hutt and S. Lotter (eds), *Epicentre to Aftermath: Rebuilding and Remembering in the Wake of Nepal's Earthquakes*, 179–202, Cambridge University Press. https://doi.org/10.1017/9781108991636.008.

Drabeck, T. E. and E. L. Quarantelli (1967), 'Scapegoats, Villains, and Disasters', *Trans-Action* 4 (4): 12–17. https://doi.org/10.1007/BF03180832.

Drèze, J. and A. Sen (1989), *Hunger and Public Action*, Oxford University Press.

Drèze, J. and A. Sen (2002), *India: Development and Participation*, Oxford University Press.

Escobar, A. (2020), *Pluriversal Politics: The Real and the Possible*, Duke University Press.

Gandhi, M. K. (n.d.), *Hind Swaraj or Indian Home Rule*. Available at: https://www.mkgandhi.org/ebks/hind_swaraj.pdf (accessed 7 November 2025).

Graeber, D. (2007), *Possibilities: Essays on Hierarchy, Rebellion, and Desire*, AK Press.

Hayward, B. and S. Johnson (2022), 'Lessons for Democracy from a Decade of Disasters', in S. Uekusa, S. Matthewman and B. C. Glavovic (eds), *A Decade of Disaster Experiences in Ōtautahi Christchurch: Critical Disaster Studies Perspectives*, 335–51, Springer Nature.

James, E. (2010), *Democratic Insecurities: Violence, Trauma, and Intervention in Haiti*, University of California Press.

Karki, T. and N. Salike (2020), 'Disaster Governance in South Asia: Special Reference to Nepal', in S. Bandyopadhyay, C. R. Pathak and T. P. Dentinho (eds), *Urbanization and Regional Sustainability in South Asia: Socio-economic Drivers, Environmental Pressures and Policy Responses*, 311–29, Springer International Publishing.

Kunreuther, L. (2018), 'Sounds of Democracy: Performance, Protest, and Political Subjectivity', *Cultural Anthropology* 33 (1): 1. https://doi.org/10.14506/ca33.1.01.

Luft, R. E. (2009), 'Beyond Disaster Exceptionalism: Social Movement Developments in New Orleans After Hurricane Katrina', *American Quarterly* 61 (3): 499–527.

Madianou, M., J. C. Ong, L. Longboan and J. S. Cornelio (2016), 'The Appearance of Accountability: Communication Technologies and Power Asymmetries in

Humanitarian Aid and Disaster Recovery', *Journal of Communication* 66 (6): 960–81. https://doi.org/10.1111/jcom.12258.

Matthewman, S. and S. Uekusa (2021), 'Theorizing Disaster Communitas', *Theory and Society* 50 (6): 965–84. https://doi.org/10.1007/s11186-021-09442-4.

National Human Rights Commission, Nepal (2020), 'The Major Highlights of the Activities of NHRC Nepal on the COVID-19', NHRC. Available at: https://nhrcnepal.org/uploads/publication/The_Major_Highlights_of_the_Activities_of_NHRC_Nepal_on_the_COVID-19,_18_May_2020.pdf (accessed 7 November 2025).

Nussbaum, M. C. (2018), *The Monarchy of Fear: A Philosopher Looks at Our Political Crisis*, Oxford University Press.

O'Grady, N. D. and Shaw (2023), 'Disaster Reparations? Rethinking Disaster Recovery Through the Politics of Affect', *Geographical Journal* 189:514–25.

O'Grady, N., D. Shaw and S. Parzniewski (2022), 'People in a Pandemic: Rethinking the Role of "Community" in Community Resilience Practices', *Geoforum* 132:32–41. https://doi.org/10.1016/j.geoforum.2022.03.021.

Oliver-Smith, A. (1999), '"What Is a Disaster?": Anthropological Perspectives on a Persistent Question', in A. Oliver-Smith and S. Hoffman (eds), *The Angry Earth: Disaster in Anthropological Perspective*, 18–35, Routledge.

Olson, R. S. (2000), 'Toward a Politics of Disaster: Losses, Values, Agendas, and Blame', *International Journal of Mass Emergencies & Disasters* 18 (2): 265–87. https://doi.org/10.1177/028072700001800201.

Onlinekhabar (2020), '17 Nepali Workers in UAE Demonstrate at Embassy Demanding Repatriation', 5 May. Available at: https://english.onlinekhabar.com/17-nepali-workers-in-uae-demonstrate-at-embassy-demanding-repatriation.html (accessed 7 November 2025).

Pelling, M. and K. Dill (2010), 'Disaster Politics: Tipping Points for Change in the Adaptation of Sociopolitical Regimes', *Progress in Human Geography* 34 (1): 21–37. https://doi.org/10.1177/0309132509105004.

'Prime Minister's Statement on Coronavirus (COVID-19): 23 March 2020' (2020), GOV.UK, 23 March. Available at: https://www.gov.uk/government/speeches/pm-address-to-the-nation-on-coronavirus-23-march-2020 (accessed 7 November 2025).

Raj, Y. and B. Gautam (2015), *Courage in Chaos: Early Rescue and Relief after the April Earthquake*, Vajra.

Rhiney, K. (2020), 'Dispossession, Disaster Capitalism and the Post-Hurricane Context in the Caribbean', *Political Geography* 78:102171. https://doi.org/10.1016/j.polgeo.2020.102171.

Sanyal, P. and V. Rao (2018), *Oral Democracy: Deliberation in Indian Village Assemblies*, Cambridge University Press.

Sen, A. (2006), *The Argumentative Indian: Writings on Indian History, Culture and Identity*, Penguin.

Siddiqi, A. (2019), *In the Wake of Disaster: Islamists, the State and a Social Contract in Pakistan*, Cambridge University Press.

Solnit, R. (2009), *A Paradise Built in Hell*, Penguin.

Staeheli, L. A. (2010), 'Political Geography: Democracy and the Disorderly Public', *Progress in Human Geography* 34 (1): 67–78. https://doi.org/10.1177/0309132509105006.

Stallings, R. (1988), 'Conflict in Natural Disasters: A Codification of Consensus and Conflict Theories', *Social Science Quarterly* 69 (3): 569–86.

Svedin, L. (2012), *Accountability in Crises and Public Trust in Governing Institutions*, Routledge.

The Kathmandu Post (2015), 'PM Calls for Unity to Stand National Crisis', 8 May. Available at: https://kathmandupost.com/miscellaneous/2015/05/08/pm-calls-for-unity-to-stand-national-crisis.

Xu, B. (2017), *The Politics of Compassion: The Sichuan Earthquake and Civic Engagement in China*, Stanford University Press.

2

Reframing 'Disasters' through Urban Political Ecology

Reflections from Two Latin American Cities

Ricardo Fuentealba and Belén Desmaison

Urban political ecology (UPE) represents one of the most interesting and innovative approaches within critical geographical literature. Geographers including David Harvey (1982), Neil Smith (1984) and Michael Watts (1983) have influenced UPE, given their Marxist standpoint on issues such as nature and labour. A more direct root of this approach relates to political ecology, a field – or 'community of practice' (Robbins 2012: 20) – that emerged from geography, development studies, environmental studies and other interdisciplinary frames to bring a more critical focus on socioecological problems. Political ecology resists the *a*political character of some ecological analyses, stressing the structural (more than a proximate one), power-laden (instead of politically inert) and normative perspective (rather than an 'objective' one) in considering human-environmental issues (Robbins 2012). However, more than the sole application of political ecology to urbanization issues, UPE emerged and developed from literatures including Marxist political economy and critical urban theory as well as science and technology studies (Swyngedouw 1996; Zimmer 2010). The approach is thus characterized by a certain 'critical predisposition' to link particular analyses of 'urban environmental problems to larger socioecological solutions', which is why UPE requires this sort of radical and critical social theory (Keil 2003: 724). During almost thirty years of UPE's trajectory, there have been multiple reviews of its development and assessments of its myriad 'waves' (Heynen 2014, 2016, 2018).

In spite of developing from some disaster-related work within political ecology, including Watts (1983), Blaikie and Brookfield (1987) and other hazards perspectives (Wescoat Jr. 2015), UPE has not been a central perspective within geographical analyses of disasters. There are exceptions, including the work of

Collins (2008), Ranganathan (2015), Marks (2015) and others. We argue that UPE can expand a more critical agenda within disaster studies by attending to the profoundly political-ecological character of disaster risk creation: that is, how particular power-laden processes of human-environmental relations remain at the root of disasters in urban centres (and beyond). More than reviewing the literature and expanding the theoretical convergences of both UPE and disaster studies (for a full review, see Desmaison and Fuentealba, forthcoming), this chapter aims to show the empirical richness of UPE as a critical geographical lens to understand a set of overlooked characteristics in traditional disaster analyses: the complex relationship of human-environmental factors, the need to transcend overly technical research, addressing the issue of knowledges and responsibilities, and approaching the everyday in risk creation.

The chapter analyses two cases in the Global South: the O'Higgins region in Chile and the Amazonian urban area of Iquitos, Peru. On the one hand, we analyse the case of O'Higgins, a region in central-south Chile characterized by an exclusive and extractivist development process that produces uneven vulnerabilities. We focus particularly on a 'rural' area heavily affected by a series of floods and geological hazards during the winter of 2023. By centring on disaster risk creation, we use this case to highlight the role of urban political features in non-city spaces, showing the relational characteristics of UPE in understanding disasters. On the other hand, in Iquitos, the Peruvian government is resettling approximately 16,000 inhabitants of the flooding neighbourhood of Belén, as it is expected that the Amazon River will overtake the area currently inhabited. We analyse state-led discourses and cartographies built to justify the decision to relocate Belenians, which showcase the imposition of a particular and dominant understanding of what it means to inhabit. We then offer alternative points of analysis and representation which highlight often overlooked but fundamental aspects of contemporary urban realities in the Amazon rainforest. In so doing, we highlight the necessity to incorporate these dimensions of inhabiting into the design and implementation of pertinent solutions to climate-induced resettlement.

The chapter is organized into the following three sections. First, we lay out four interconnected dimensions that show how a UPE lens can be integrated into disaster studies: UPE moves analyses of disasters beyond dualism; it centres the importance of politics in the (re)creation of urban disasters; it recognizes the role of knowledge, power and responsibility in risk creation and finally, it centres on the role of 'everyday', 'slow' disasters and situated politics. Next, we expand upon the two empirical cases, applying the four dimensions to demonstrate how UPE can be deployed to advance the analysis of disasters. We conclude the chapter by discussing the commonalities of the cases as well as where UPE and disaster studies might go.

Four Dimensions to Operationalize UPE for Disasters

It is commonly recognized that disasters result from the encounter between human and natural phenomena or, in the field's terminology, vulnerability and hazards (Wisner et al. 2004). Such a binary, however, has been criticized lately for limiting disaster understanding and excluding non-Western perspectives (Gaillard 2021). Within such an agenda, UPE works to transcend reductive dualistic thinking, particularly the human/nature divide, in disaster research. This is not a unique feature of UPE, as post-humanist perspectives, including assemblages and more-than-human geographies, have also addressed this relationality (McGowran and Donovan 2021; van Riet 2021). UPE, however, addresses this hybridity between socio-natural entities as a way to show how the city, for example, is ultimately *at the same time* nature and society (Swyngedouw 1996). It stems from Bruno Latour's (1993) analysis of modern science. This modernist separation manifests as the binary of hazards and vulnerability, encapsulated in the formula Risk = Vulnerability × Hazard (Wisner et al. 2004).

Failing to address the hybrid socioecological processes that (re)produce vulnerability, we argue, perpetuates the oversimplification and depoliticization of disaster risk management. This results in overly technical solutions that, on the one hand, overlook people's ability to cope and act and, on the other hand, fail to address the root causes of vulnerability and risk. UPE contributes by using concepts such as metabolism (i.e. the material production of socio-nature (see Swyngedouw 1996, 2006)) to overcome this dualism, arguing that environmental and social changes are interconnected and mutually influence each other. Hence, socio-spatial conditions are shaped not only by social, political and economic factors but also by physical, chemical and biological processes (Heynen et al. 2006). In disaster studies, this requires recognizing complex socioecological processes instead of static demographics: that is, instead of inferring that particular groups are vulnerable due to their class or gender (Jacobs 2019; Rivera et al. 2021), their disaster experiences result from complex histories that include such socioecological processes. Overcoming the city/nature divide involves, for example, adopting a new ontology that values the interdependence of humans and more-than-human entities, as seen in Amazonian settlements (Kohn 2013). Recognizing diverse world-views, or 'pluriverses' (Escobar 2020), can transform disaster risk management and climate change adaptation by promoting sustainability and harmony between nature and human activity.

Second, mainstream disaster studies have a tendency to avoid critical questions about power, leading to top-down technical solutions that overlook local realities and emphasize engineering and physical interventions (Watts 1983). This ends up perpetuating risks by not challenging the socio-political aspects that reproduce them (Gaillard 2019). This leads to a focus on proximate causes (i.e. natural, 'local' or less spatially distant factors) and material impacts that treat *symptoms* rather than *root causes*, which are deeply rooted in political and economic systems (Gibb 2018). Disasters are seen as immediate events rather than processes that build over time, obscuring structural causes of increased hazard exposure. Against these standpoints, UPE can be a helpful antidote to reductive analyses because it helps to understand the broader processes leading to disasters and challenge underlying issues for more just outcomes. UPE highlights the social and political production of risks, showing how terms like 'resilience' and 'vulnerability' become depoliticized if they do not engage with how risk is produced, shifting responsibility away from the state and obscuring citizen agency (Marks 2015; Bankoff 2019; von Meding 2021). UPE's acknowledgement of socioecological processes as both material and discursive can help to approach vulnerability as a socially constructed label imposed on marginalized groups, legitimizing top-down disaster management and depicting these groups as helpless and in need of external assistance (Jacobs 2019; Lenette 2019; Chmutina et al. 2022). This reflects unequal power relations in decision-making about who is considered vulnerable or resilient and whose knowledge is valued.

Third, UPE helps to approach different types of knowledge/power and their role in creating and managing at-risk conditions. As the social and the environmental co-constitute each other, UPE recognizes the discursive dimension of 'knowing' nature through certain representational systems and discursive apparatuses (Swyngedouw 1996; Heynen et al. 2006). In combination with other literatures such as feminist and postcolonial political ecologies (see Heynen 2016, 2018; Chatti 2025), UPE can question how certain ways of knowing and producing nature come to exist, who is included and excluded, and why. In disaster research, this means questioning who is represented in research and the purposes behind it, including global asymmetries of knowledge production (Gaillard 2019). This relates to what Fricker (2013) calls epistemic injustice, that is, how certain forms of knowledge are discredited and thus can perpetuate structural drivers of risk and vulnerability. Decolonial literature advocates for recognizing and amplifying diverse knowledges and world-views, arguing for the coexistence of multiple pluriverses (Escobar 2020). Contemporary UPE literature includes reconsidering the universality of overarching environmental discourses, including the Anthropocene (Ernstson and Swyngedouw 2019) and the climate emergency (Kaika et al. 2023). This work argues against the notion that all humans are equally responsible for climate change, attributing the crisis to the inherent drive for capital accumulation and economic exploitation by elite power groups (Demos 2017; Moore 2017). Along this line, UPE also helps to understand how and why impacts and adaptive capacities to crises are unequally distributed, necessitating the inclusion and amplification of historically marginalized knowledge in decision-making processes for urbanization and effective disaster risk management (Escobar 2020). An emerging argument in climate change discussions, such as COP meetings, revolves around 'loss and damage', with less polluting countries arguing that major polluters should bear more financial responsibility for future climate change impacts (Mechler et al. 2018).

Finally, within disaster studies, UPE can help to shift attention towards the 'everyday' and 'slow' aspects of risks and disasters. Disaster researchers have criticized the overarching emphasis on major disaster events, advocating for an understanding of how risks accumulate through everyday hazards in urban areas (Bull-Kamanga et al. 2003). This leads to incorporating lived experiences into scholarly and policy-oriented knowledge to challenge superficial views and recognize diverse ways of being and understanding (Amin and Lancione 2022). Lawhon et al. (2014) introduce 'situated urban political ecology', which expands traditional UPE by incorporating African urbanism and focusing on everyday practices and diffuse power in urbanization. This approach, described as 'radical incrementalism', emphasizes gradual changes rather than systemic overhauls. Methodologically, it advocates for ethnographic research to understand environments within their specific contexts (Gupta and Sharma 2006). From this perspective, UPE also critiques the concept of disasters as spectacular events, proposing instead the examination of less visible, gradual harms akin to 'slow violence' (Nixon 2011) and 'slow disasters' (Knowles 2014). These include long-term environmental degradation and deferred maintenance, where impacts unfold over generations rather than as immediate, catastrophic events. Recognizing these slow-moving disasters allows for a more nuanced understanding of environmental activism and local resistance, highlighting the political nature of socioecological processes and risk construction (Ingram et al. 2006).

The field of UPE has a long tradition and faces multiple challenges associated with urban environmental issues (Tzaninis et al. 2021). We have only focused on four dimensions that, we think, help to integrate UPE into disaster studies. This aims to enhance our understanding of complex lived realities and the social dimensions of risk production towards the production of more pertinent solutions to disaster risk reduction. In the following two sections, we operationalize these four dimensions in two different but complementary case studies.

The Urban Political Roots of Disasters in Rural Chile: The Case of O'Higgins

This section centres on the O'Higgins region in central-south Chile, where in 2023 two major meteorological events triggered floods and mass-wasting processes that affected both city and non-city areas. Local communities living in the Cachapoal Valley interpreted the meteorological events as extremely rare. After years of drought in the region, including a number of *comunas* (districts) with water scarcity warnings, the June 2023 rain was not 'normal', nor was that of August 2023. However, while these hazardous events might not have been expected, the myriad processes of risk creation can be excavated, thanks to UPE. What occurred in 2023 results from a long history of socioecological processes in the region – processes which must be understood by transcending traditional disaster dualisms of hazards and vulnerabilities, as UPE encourages. In concrete terms, the root causes of the 2023 disasters go beyond two types of binaries: the nature/culture dichotomy and the urban/non-urban distinction. This can be seen with two examples.

First, the communities of O'Higgins have a strong dependency on both human and non-human elements, given their rural livelihoods. A central characteristic of regional economic development is the coexistence of smallholder agriculture alongside industrial single-crop farming for urban consumption and exports. The region as a whole – as in Chile – is not a poor territory but an unequal one: it is characterized by a rich environment with multiple resources but unevenly distributed. In the face of the 2023 meteorological events, agroindustrial farms that are well connected with global capital and urban markets did not suffer major losses. This was not the case for smallholder agriculture and low-income groups. The disaster can thus only be understood when a geographically uneven trajectory is considered, as it results from a set of socioecological and material practices within capitalist-led agroindustrial development. And second, the role of the double duality of nature/culture and urban/non-urban processes in creating disaster risks also relates to a particular economic activity that locals recognize as a basis for risk creation: aggregate extraction. Aggregate extraction refers to the metabolic practice of removing material from riverbeds and estuaries and transforming it into commodities such as stone inputs, which are later used for infrastructure development and construction. This activity, which is poorly regulated in local *comunas*, entails the literal extraction and circulation of flows from non-urban to urban areas. Such a metabolic urbanization (Swyngedouw 2006) is thus by definition a relational spatial process that links city and non-city areas. And as acknowledged in some local accounts, aggregate extraction has adverse impacts on environmental degradation and local livelihoods, increasing the exposure of local communities to hazards and exacerbating their disaster vulnerabilities.

A second contribution we obtain from UPE refers to the politics of (urban) disasters. Traditional and dominant explanations of the O'Higgins disasters centre on proximate causes such as flood hazards, housing materials and their location, or the absence of emergency warnings. Many of these *a*political causations therefore argue that the unexpectedness of the hazard event, including its intensity within mountainous areas, is what explains the disaster. Disasters, however, are not natural. Given the extent of the impacts, with urban and rural spaces flooded, hundreds of people isolated and critical damage to infrastructure, this makes sense. Centring on the characteristics of the hazards and their physical impacts is underpinned by a technocratic or naturalistic interpretation of disaster processes which obscure the ingrained socioecological relations that remain at the root of disasters. A UPE-inspired analysis of the floods in O'Higgins shows the extent to which social and ecological questions are intimately ingrained in the politics of risk creation.

An important dimension of this relates to the aforementioned long-standing process of uneven development. The most impacted area in the Cachapoal Valley of O'Higgins shows the coexistence of two groups: smallholder farmers and an agroindustrial sector. The former possess scarce assets, including land and water rights, while the latter benefits from historical and institutional settings that support export-driven rural development. Following UPE, the geographically uneven character of the region reflects the socio-environmental conditions that enable the production and circulation of particular agricultural commodities for export. Therefore, according to a UPE analysis, the root causes of vulnerabilities in O'Higgins relate to both material and discursive elements associated with food production for urban and international markets in rural spaces. From this process, however, local communities understand resilience and vulnerability as profoundly political. We find that they explicitly recognize that their resilience and vulnerability to disasters are shaped by processes that go 'beyond nature', focusing instead on extractive agricultural activities and their politico-institutional counterpart. They mention economic and export-driven agricultural development, as well as the exclusionary water management norms, as the causes. Mobilized communities criticize the extractive economic activities in place, given their social and environmental impacts. This includes the concentration of industrial pig farming and the usage of agrochemical components for food production. It is in the face of these exclusionary economic activities that locals understand their vulnerable conditions and disaster experiences: they are *rendered* vulnerable and powerless. That is why they stress the need for more participatory processes of land use management that can redirect the trajectory of the territory and transform the development of the region.

A third dimension relates to knowledge/power and responsibility in risk creation, which goes hand in hand with the exclusionary development process in place in O'Higgins. Dominant interpretations of disasters focus on technical aspects of hazardous events, as seen in media, local bureaucracies and scientific assessments. In this mainstream view, locals' voices are reduced to suffering, occupying a passive role in the naturalized interpretation. UPE, on the contrary, given its emphasis on how contesting knowledges exist in (re)creating disaster risks, acknowledges the myriad ways in which particular discourses are constructed as 'expert' over others, given their entanglement with power. This recognition thus leads to a politically laden and denaturalized idea of responsibilities in risk creation.

Territorial conditions in O'Higgins, including its risks, result from the actions of actors and institutions following an extractive, unsustainable and pro-economic-growth development model. This governance arrangement includes how big landowners and private companies influence and legitimize this particular way of developing rural spaces as dominant, which is the result of a neoliberal agenda imposed during the Chilean dictatorship (Harvey

2005). This arrangement benefits a particular elite at the expense of the wider group of smallholder farmers, who also intend to participate in local and regional markets but are unable to compete. Additionally, neoliberal policy processes have scarcely included local knowledges and experiences of these smaller, non-elite groups. This resulted in decades of disaster risk management based on command-and-control perspectives, focused only on emergencies, and failing to hold accountable the actors responsible for (re)creating risks (Saavedra 2021). In such an arrangement, communities occupy a secondary position. As argued by Bustos and Román (2024), while rural communities in Chile have been greatly affected by neoliberal policies, their importance and representation in formal political processes remain secondary. This has limited the number and quality of spaces through which these communities exercise their democratic participation and construct their citizenship. This includes rural and environmental mobilization in territories characterized by dominant extractive economies such as O'Higgins. That is why local collective action is using a more diverse repertoire of knowledges and strategies to influence their political contexts, which remains to be seen in how effectively it transforms local governance (Martínez and Maillet 2024).

And finally, the fourth dimension relates to everyday environments and situated politics. The 'unexpectedness' of the June and August 2023 floods has already been stated. From a detached view, this has been interpreted as the unfortunate impacts of nature and/or the effects of climate change. A UPE-oriented analysis centred on the everyday, however, focuses on how risks are the result of an accumulative socioecological process. There are two facets of this: on the one hand, how smaller events of rainfall-related floods have occurred, although with less spectacular impacts. That is, how the trajectory of O'Higgins, with its uneven socioecological distribution of resources and opportunities, has produced smaller but more frequent disruptions. And on the other, how floods and geological hazards interact with other types of hazards. The latter is particularly important because, as previously mentioned, the development of O'Higgins has been marked by years of drought. Therefore, a complex and steady geographical trajectory of this area, which relates to a long-standing development that slowly incubated local vulnerabilities and risks, remains at the core of what happened in 2023. In other words, the socio-environmental arrangement leading up to the 2023 floods can only be understood if the everyday accumulation of risks, as an expression of the dominant development path of the region, is considered.

It is in that context that local communities have sought to develop collective actions towards inclusion and conservation. These mobilizations represent a type of situated politics that entwines the agendas of social and environmental justice in the context of at-risk territories. For instance, in the context of droughts and an uneven distribution of water rights, local organizations and activists have raised the alarm for years regarding water usage in the territory. They have organized multiple events and encounters with political representatives and regional authorities, with the aim of regaining sovereignty over natural resources. More recently, the same has occurred with some conflictive cases of aggregate extraction, as environmental activists are going to trials over environmental damage due to this land usage. Therein, their strategies are diverse and contextual, typically involving mobilizations in cities – where judicial bureaucracies are located. Ultimately, these practices represent a particularly situated contestation of the socioecological risks present in O'Higgins.

Redressing Displacement with Counter-representations of Everyday Places in the Peruvian Amazonia

From its early days, the Amazonian city of Iquitos, Peru, has had a relationship of interdependence with the self-built settlement of Belén, which provides essential services and operates the largest market in the Peruvian Amazon. Belén was formed by indigenous families who migrated to the city in search of jobs and better education for younger generations. From its inception, Iquitos was a dual city divided between the 'formal' European-like areas and the self-built settlements made with local materials (wood, palm trees) and vernacular architectural techniques.

Showcasing ontologies that challenge dualistic understandings such as the human/nature divide, people in Amazonian settlements engage with water in symbolic and productive terms. From a very early age, they refer to themselves as 'amphibious societies', for which rivers are the main means of mobility, livelihoods, and symbolic and spiritual significance. The relationship between Belenians, the river and the forest is explained by Padoch et al. (2008) as 'multi-situatedness' in urban forests or rural cities. Belenians are mobile people for whom the limits between rural and urban are diffused. People are in constant flux, and the number of people living in one household varies throughout the year, as some go into the forest during crop and fishing seasons, and rural dwellers visit the city temporarily for health, education and other services.

Life in the Amazon rainforest is dictated by the cyclical rhythms of nature and society. The changing and complex nature of amphibian settlements between flooding and dry seasons (Figure 2.1) alters not only people's movement through space but also space itself: a built environment dictated by temporality, the makeshift, constant renewal, expansion and contraction. Yet these rhythms are absent in current urban planning and state cartographies (Figure 2.2).

The dynamism of the territory, with its flooding seasons and continuously changing course of rivers (Figure 2.3), results in tensions with sedentary and ever-expanding settlements that follow urban and architectural models developed elsewhere. The resettlement project of Nueva Ciudad de Belén (NCB) (Figure 2.4), promoted by the Peruvian government, is the latest manifestation of an imposition of a particular vision of what inhabiting entails, which obliterates other understandings of the relationships between humans and nature. What lessons do amphibian settlements provide for the urgent transformation required for Amazonian urbanism to face the challenges of the current climatic crisis?

Figure 2.1 A street in Belén during the flood season.
Source: Drawing by Belén Desmaison and Reynaldo Siccha based on photographs, field notes, sketches and interviews with Belenians, 2024.

Figure 2.2 Lower area of Belén as represented in state cartography.
Source: Produced by Belén Desmaison and Reynaldo Siccha based on data provided by the Ministry of Housing, Construction and Sanitation of Peru, 2024.

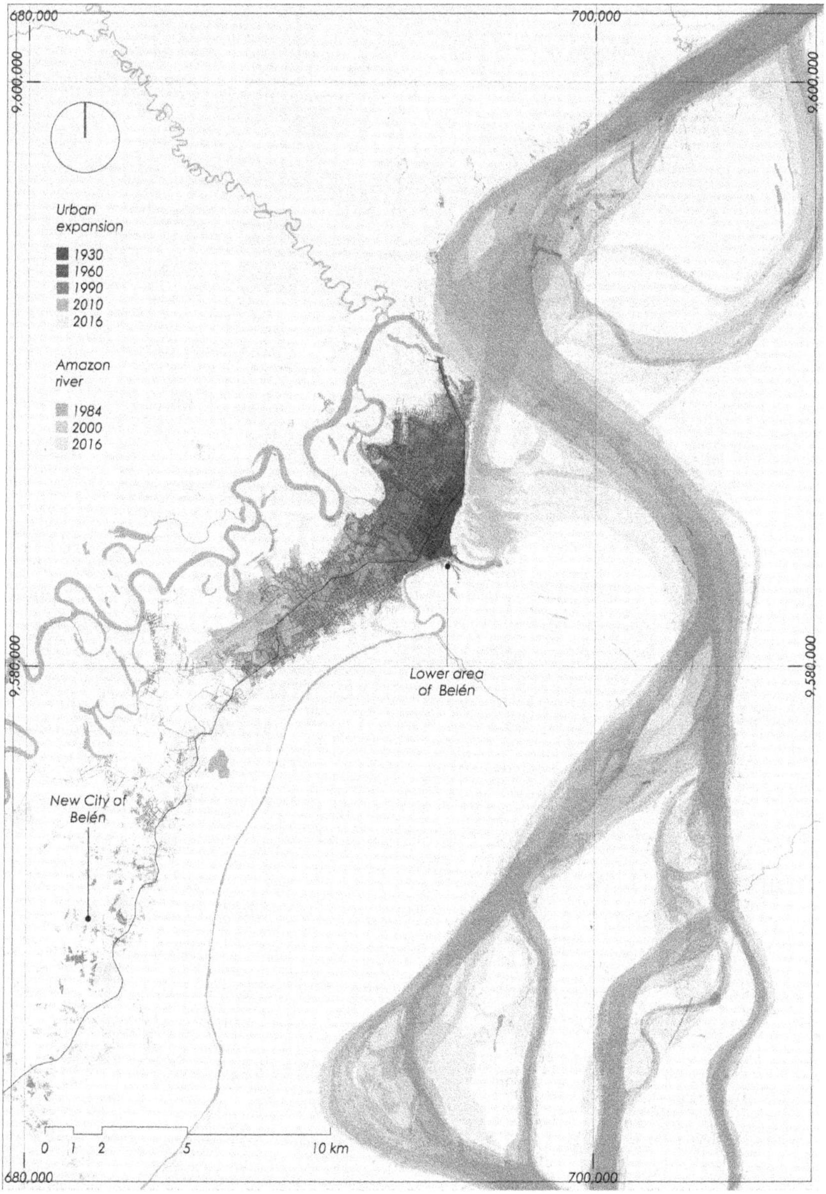

Figure 2.3 Iquitos, Peru.
Source: Drawing by Belén Desmaison and Reynaldo Siccha based on aerial photographs, 2024.

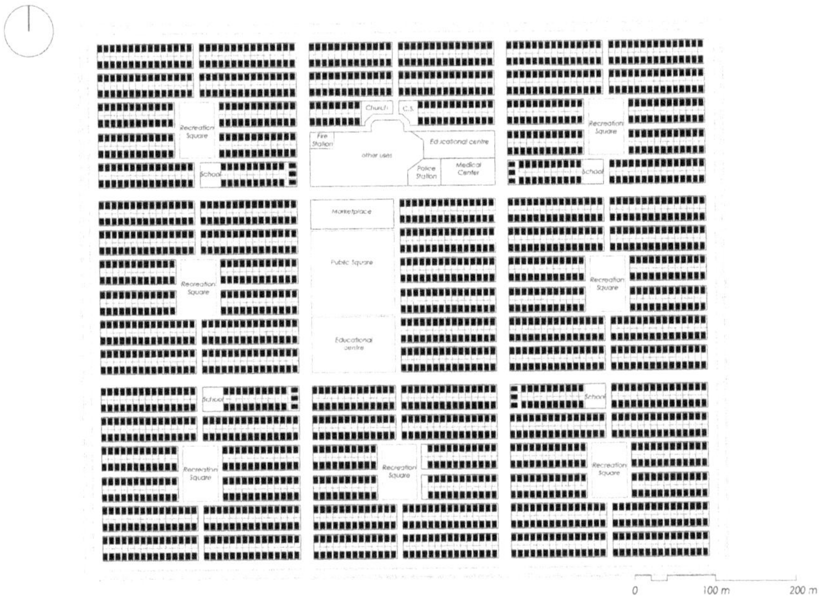

Figure 2.4 The project of Nueva Ciudad de Belén as represented in state-led cartography. *Source: Produced by Belén Desmaison and Reynaldo Siccha based on data provided by the Ministry of Housing, Construction and Sanitation of Peru, 2024.*

In Peru, state-led approaches to disaster risk management and planning continue to emphasize technical and quantitative factors, rendering invisible the socio-political dimensions that reproduce vulnerability and risk. This situation is graphically reproduced in cartographies produced by national and international institutions, which 'present themselves as objective and value-free', overwriting 'historical conditions and internal heterogeneity, providing a tabula rasa for the operation of state power and capitalist social relations' (Anthias 2019: 225–6). Countering this approach entails addressing and understanding 'the agency of visual data in contemporary discourse and world making, namely, what is made visible, what is overlooked or not seen, where agency lies' (Troiani and Ewing 2021: 8).

Power imbalances are clear in the Peruvian case, where the state operates within policy frameworks that tend to render the agency and voices of the displaced invisible in decision-making processes. Li (2005: 384), following Scott (1998), explains how government representatives position themselves 'as experts who know how others should live, they collect and arrange data according to simplified grids, they diagnose deficiencies, and they devise elaborate interventions to bring about improvement'. The state uses different instruments such as 'cadastral maps, housing types or building codes to make society "legible", simplified and standardised for easy functioning' (Johnson et al. 2021: 86). The cartographies of Belén and the projected NCB depict settlements with spatial layouts that obliterate identity and diversity.

The proposed 'solution' to the imminent threat of flooding does not adapt to the everyday realities of those displaced, limiting the possibilities of creating homes and places, and reducing complex realities to 'bare life' in search of control and order. The prefix 'dis' implies being apart or away, creating a distance between the subject and 'place'. In that sense, displacement can be understood as the loss of social, environmental and emotional connections, as well as a sense of place. Displacement is the negation of 'place' and home, and NCB has done so both in terms of negating Amazonian urban amphibian inhabitation and in taking away the forest on which livelihoods depend.

Risks are socially constructed and increase over time. Prior to the accelerated urban growth of the mid-twentieth century onwards, amphibian Amazonian settlements adapted to flood cycles and to the changing course of rivers by temporarily inhabiting areas, building temporary houses and undergoing constant resettling (Figure 2.5). At the turn of the twenty-first century, flooding became a 'risk' due to changing settlement patterns in rapidly expanding self-built settlements like Belén. The accelerated urban growth of Iquitos and other cities in Peru can be partially attributed to changing conceptions of property promoted by the state. 'Property' came to be understood as individually owned sites with clearly delimited borders, which became key to accessing public services.

The aspiration for permanence and stability in a territory that is continuously eroded and flooded is contradictory. This desire and settlement pattern is materially expressed in the shift from floating houses to houses on stilts built with durable materials like concrete. This material shift results in risks, as infrastructures that are permanently in place are not able to withstand constant flooding or rapid erosion resulting from the changing course of rivers.

In the case of Belén, we aim to understand how risk accumulates through everyday socio-natural hazards not only through discourse and writing but also through alternative representational and analytical cartographies and drawings. In the search for ways to unveil emotions, senses and everyday experiences, visual research methods are 'not purely illustrative. They also contribute to the process of documentation; hence they are communicative yet analytical, revealing patterns of everyday life that are often missed or hidden' (Sharif 2017: 11).

Figure 2.5 Belén during the flood and dry seasons.
Source: Drawing by Belén Desmaison and Reynaldo Siccha based on photographs, field notes, sketches and interviews with Belenians, 2024.

Conclusion

Geography is undoubtedly one of the core disciplines within disaster studies. However, more critical perspectives within and beyond this discipline are required. UPE, as we have shown, represents an important perspective for critically researching and acting upon the myriad disaster processes. Using cases from Chile and Peru, we have applied a set of dimensions inspired by UPE, from which we can extract a number of lessons.

A UPE focus on disaster risk management is useful as it highlights often overlooked interdependencies between social, political, economic and natural processes. State-led narratives tend to oversimplify complex socio-natural processes by focusing solely on the natural dimensions of risks, rendering invisible the social dimensions that reproduce inequalities and risk exposure, hence limiting people's capacities to overcome risk. Moreover, this simplistic view is also manifested in state-led cartographies and representations that depict overly technical territories where everyday realities are absent.

Binary and sectorial approaches to complex lived realities result in solutions that do not tackle the causes of differentiated risk exposure. As seen in both case studies, reductionism and the oversimplification of risk result in the reproduction of social and economic drivers of risk and inequality. Recognizing the unsustainability and injustice of current approaches to disaster risk management, we argue that UPE is a useful lens for moving beyond dualisms towards more pertinent solutions able to tackle multidimensional drivers of risk.

The cases also emphasize how local knowledges are marginalized, making it difficult for local communities to exercise their agency in the face of dominant power structures. In O'Higgins, the dominant paradigm of economic growth does not consider rural communities, reducing their participation and failing to hold accountable those responsible for risk creation. Similarly, in the case of the displacement of Belenians, there is a loss of 'place' and identity that is enforced by top-down policies that end up stripping their agency. Our UPE analysis then shows the need to recognize diverse knowledges that can foster inclusive governance to truly address complex disaster processes.

Finally, the cases explore the role of everyday practices and situated politics in disaster processes. This is shown through the accumulation of socioecological processes tied to dominant development paths, as in O'Higgins, with responses from local environmental activists intending to influence such paths. In Iquitos, everyday practices are understood as the ordinary and banal events that are often overlooked in planning but which carry emotional and symbolic importance in creating a bond between people and 'place'. The drawings of Belén seek to emphasize these invisible (in planning) events to contest top-down climate adaptation strategies. This grounded approach within UPE helps reconsider the overly structural changes associated with Marxist-UPE and expands upon the role of everyday practices and different ways of knowing in incrementally transforming at-risk areas.

Acknowledgements

Ricardo acknowledges support from ANID/Fondecyt Research Starting Grant Project #11240523. Belén acknowledges the support of the Durham Doctoral Studentship and PUCP's funding for creative research (CAP Creación PI1206).

References

Amin, A. and M. Lancione (eds) (2022), *Grammars of the Urban Ground*, Duke University Press.

Anthias, P. (2019), 'Ambivalent Cartographies: Exploring the Legacies of Indigenous Land Titling Through Participatory Mapping', *Critique of Anthropology* 39 (2): 222–42. https://doi.org/10.1177/0308275X19842920.

Bankoff, G. (2019), 'Remaking the World in Our Own Image: Vulnerability Resilience and Adaptation as Historical Discourses', *Disasters* 43 (2): 221–39. https://doi.org/10.1111/disa.12312.

Blaikie, P. and H. Brookfield (1987), *Land Degradation and Society*, Routledge.

Bull-Kamanga, L., K. Diagne, A. Lavell, E. Leon, F. Lerise, H. MacGregor, A. Maskrey, M. Meshack, M. Pelling, H. Reid, D. Satterthwaite, J. Songsore, K. Westgate and A. Yitambe (2003), 'From Everyday Hazards to Disasters: The Accumulation of Risk in Urban Areas', *Environment and Urbanization* 15 (1): 193–204. https://doi.org/10.1177/095624780301500109.

Bustos, B. and A. Román (2024), 'Prácticas ciudadanas en territorios rurales: efectos de 40 años de neoliberalización económica', *Revista de Geografía Norte Grande* 87:1–23. Available at: https://revistanortegrande.uc.cl/index.php/RGNG/article/view/41977 (accessed 7 November 2025).

Chatti, D. (2025), 'Gendering Nature: From Ecofeminism to Feminist Political Ecology', in G. L. Simon and K. Kay (eds), *Doing Political Ecology*, 223–37, Routledge.

Chmutina, K., J. von Meding, D. Williams, J. Vickery and C. Purdum (2022), 'From Pity to Fear: Security as a Mechanism for (Re)production of Vulnerability', *Disasters* 47 (3): 546–62. https://doi.org/10.1111/disa.12568.

Collins, T. W. (2008), 'The Political Ecology of Hazard Vulnerability: Marginalization, Facilitation and the Production of Differential Risk to Urban Wildfires in Arizona's White Mountains', *Journal of Political Ecology* 15:21–43. https://doi.org/10.2458/v15i1.21686.

Demos, T. J. (2017), *Against the Anthropocene: Visual Culture and Environment Today*, Sternberg Press.

Desmaison, B. and R. Fuentealba (Forthcoming), 'Urban Political Ecology and the Reproduction of Disasters', in K. Chmutina, R. Fuentealba and D. Rivera (eds), *Principles and Concepts in Disaster Risk: Reducing Risks: A WSPC Reference on Preventing and Mitigating Disasters*, vol. 1, World Scientific Publishing.

Ernstson, H. and E. Swyngedouw (2019), *Urban Political Ecology in the Anthropo-Obscene: Interruptions and Possibilities*, Routledge.

Escobar, A. (2020), *Pluriversal Politics: The Real and the Possible*, Duke University Press. https://doi.org/10.1515/9781478012108.

Fricker, M. (2013), 'Epistemic Justice as a Condition of Political Freedom?', *Synthese* 190 (7): 1317–32. https://doi.org/10.1007/s11229-012-0227-3.

Gaillard, J. C. (2019), 'Disaster Studies Inside Out', *Disasters* 43 (S1): S7–S17. https://doi.org/10.1111/disa.12323.

Gaillard, J. C. (2021), *The Invention of Disaster: Power and Knowledge in Discourses on Hazard and Vulnerability*, Routledge.

Gibb, C. (2018), 'A Critical Analysis of Vulnerability', *International Journal of Disaster Risk Reduction* 28 (June): 327–34. https://doi.org/10.1016/j.ijdrr.2017.11.007.

Gupta, A. and A. Sharma (2006), 'Introduction: Rethinking Theories of the State in an Age of Globalization', in A. Gupta and A. Sharma (eds), *The Anthropology of the State: A Reader*, 1–41, Wiley-Blackwell.

Harvey, D. (1982), *The Limits to Capital*, Basil Blackwell.

Harvey, D. (2005), *A Brief History of Neoliberalism*, Oxford University Press.

Heynen, N. (2014), 'Urban Political Ecology I: The Urban Century', *Progress in Human Geography* 38 (4): 598–604.

Heynen, N. (2016), 'Urban Political Ecology II: The Abolitionist Century', *Progress in Human Geography* 40 (6): 839–45.

Heynen, N. (2018), 'Urban Political Ecology III: The Feminist and Queer Century', *Progress in Human Geography* 42 (3): 446–52. https://doi.org/10.1177/0309132517693336.

Heynen, N., M. Kaika and E. Swyngedouw (eds) (2006), *In the Nature of Cities: Urban Political Ecology and the Politics of Urban Metabolism*, Routledge.

Ingram, J. C., G. Franco, C. R. del Rio and B. Khazai (2006), 'Post-Disaster Recovery Dilemmas: Challenges in Balancing Short-Term and Long-Term Needs for Vulnerability Reduction', *Environmental Science and Policy* 9 (7–8): 607–13. https://doi.org/10.1016/j.envsci.2006.07.006.

Jacobs, F. (2019), 'Black Feminism and Radical Planning: New Directions for Disaster Planning Research', *Planning Theory* 18 (1): 24–39. https://doi.org/10.1177/1473095218763221.

Johnson, C., G. Jain, V. Nalla and J. D. Cáceres-Martínez (2021), 'Risk as a Subjective Concept and Its Influence on Decision-Making', in C. Johnson, G. Jain and A. Lavell (eds), *Rethinking Urban Risk and Resettlement in the Global South*, 77–95, UCL Press.

Kaika, M., R. Keil, T. Mandler and Y. Tzaninis (2023), *Turning Up the Heat: Urban Political Ecology for a Climate Emergency*, Manchester University Press.

Keil, R. (2003), 'Urban Political Ecology', *Urban Geography* 24 (8): 723–38.

Knowles, S. G. (2014), 'Learning from Disaster?: The History of Technology and the Future of Disaster Research', *Technology and Culture* 55 (4): 773–84. https://doi.org/10.1353/tech.2014.0110.

Kohn, E. (2013), *How Forests Think: Toward an Anthropology Beyond the Human*, University of California Press.

Latour, B. (1993), *We Have Never Been Modern*, Harvard University Press.

Lawhon, M., H. Ernstson and J. Silver (2014), 'Provincializing Urban Political Ecology: Towards a Situated UPE Through African Urbanism', *Antipode* 46 (2): 497–516. https://doi.org/10.1111/anti.12051.

Lenette, C. (2019), *Arts-Based Methods in Refugee Research: Creating Sanctuary*, Springer Singapore. https://doi.org/10.1007/978-981-13-8008-2.

Li, T. M. (2005), 'Beyond "the State" and Failed Schemes', *American Anthropologist* 107 (3): 383–94.

Marks, D. (2015), 'The Urban Political Ecology of the 2011 Floods in Bangkok: The Creation of Uneven Vulnerabilities', *Pacific Affairs* 88 (3): 623–51. https://doi.org/10.5509/2015883623.

Martínez, F. and A. Maillet (2024), 'La articulación regional de la movilización ambiental en Chile: el caso de la Asociación Regional Ambiental de O'Higgins (ARAO)', *EURE* 51 (152): 1–26. https://doi.org/10.7764/EURE.51.152.04.

McGowran, P. and A. Donovan (2021), 'Assemblage Theory and Disaster Risk Management', *Progress in Human Geography* 45 (6): 1601–24. https://doi.org/10.1177/03091325211003328.

Mechler, R., L. Bouwer, T. Schinko, S. Surminski and J. Linnerooth-Bayer (eds) (2018), *Loss and Damage from Climate Change*, Springer Berlin Heidelberg.

Moore, J. (2017), 'The Capitalocene, Part I: On the Nature and Origins of Our Ecological Crisis', *Journal of Peasant Studies* 44 (3): 594–630. https://doi.org/10.1080/03066150.2016.1235036.

Nixon, R. (2011), *Slow Violence and the Environmentalism of the Poor*, Harvard University Press.

Padoch, C., E. Brondizio, S. Costa, M. Pinedo-Vasquez, R. R. Sears and A. Siqueira (2008), 'Urban Forest and Rural Cities: Multi-sited Households, Consumption Patterns, and Forest Resources in Amazonia', *Ecology and Society* 13 (2): 1–16.

Ranganathan, M. (2015), 'Storm Drains as Assemblages: The Political Ecology of Flood Risk in Post-Colonial Bangalore', *Antipode* 47 (5): 1300–20. https://doi.org/10.1111/anti.12149.

Rivera, D. Z., B. Jenkins and R. Randolph (2021), 'Procedural Vulnerability and Its Effects on Equitable Post-Disaster Recovery in Low-Income Communities', *Journal of the American Planning Association* 88 (2): 220–31. https://doi.org/10.1080/01944363.2021.1929417.

Robbins, P. (2012), *Political Ecology: A Critical Introduction*, Wiley-Blackwell.

Saavedra, J. (2021), 'Gobierno del desastre en el neoliberalismo: Una mirada genealógica desde los terremotos en Chile durante el siglo XX', *Revista de Estudios Latinoamericanos sobre Reducción del Riesgo de Desastres REDER* 5 (2): 48–62. Available at: https://revistareder.com/handle-0719-8477-2021-088 (accessed 7 November 2025).

Scott, J. C. (1998), *Seeing Like a State: How Certain Schemes to Improve the Human Condition Have Failed*, Yale University Press.

Sharif, Y. (2017), *Architecture of Resistance: Cultivating Moments of Possibility within the Palestinian/Israeli Conflict*, Routledge.

Smith, N. (1984), *Uneven Development: Nature, Capital, and the Production of Space*, Blackwell.

Swyngedouw, E. (1996), 'The City as a Hybrid: On Nature, Society and Cyborg Urbanization', *Capitalism, Nature, Socialism* 7 (2): 65–80. https://doi.org/10.1080/10455759609358679.

Swyngedouw, E. (2006), 'Metabolic Urbanization: The Making of Cyborg Cities', in N. Heynen, M. Kaika and E. Swyngedouw (eds), *In the Nature of Cities: Urban Political Ecology and the Politics of Urban Metabolism*, 20–39, Routledge.

Troiani, I. and S. Ewing (eds) (2021), *Visual Research Methods in Architecture*, Intellect.

Tzaninis, Y., T. Mandler, M. Kaika and R. Keil (2021), 'Moving Urban Political Ecology Beyond the "Urbanization of Nature"', *Progress in Human Geography* 45 (2): 229–52. https://doi.org/10.1177/0309132520903350.

van Riet, G. (2021), 'The Nature–Culture Distinction in Disaster Studies: The Recent Petition for Reform as an Opportunity for New Thinking?', *International Journal of Disaster Risk Science* 12:240–49. https://doi.org/10.1007/s13753-021-00329-7.

von Meding, J. (2021), 'Reframing Vulnerability as a Condition of Potential', *Arrow* 8 (1). Available at: https://arrow-journal.org/reframing-vulnerability-as-a-condition-of-potential/ (accessed 7 November 2025).

Watts, M. (1983), 'On the Poverty of Theory: Natural Hazards Reviewed in Context', in K. Hewitt (ed.), *Interpretations of Calamity: From the Viewpoint of Human Ecology*, 231–62, Allen & Unwin.

Wescoat, J. L., Jr. (2015), 'Political Ecology of Risk, Hazards, Vulnerability, and Capacities', in T. Perreault, G. Bridge and J. McCarthy (eds), *The Routledge Handbook of Political Ecology*, 293–302, Routledge.

Wisner, B., P. Blaikie, T. Cannon and I. Davis (2004), *At Risk: Natural Hazards, People's Vulnerability and Disasters*, Routledge.

Zimmer, A. (2010), 'Urban Political Ecology: Theoretical Concepts, Challenges, and Suggested Future Directions', *Erkunde* 4:343–54.

3

The Plural Lives of Rubble

A Research Agenda for Disaster Geographies

Giovanna Gioli and Amitangshu Acharya

Searching Lives in the Rubble

One might fairly ask, Was the Enlightenment buried in the rubble of Gaza, or was the rubble the Enlightenment's logical conclusion?

– Joe Sacco, 'War on Gaza', 2024

The rubble is alive and singing.

– Alison Phipps, 'Rubble', 2025

As we write, in early 2025, Gaza has been disassembled into a graveyard of 40 million tonnes of rubble. According to reports from various UN agencies, of the 50,000 and more who have been killed, at least 10,000 corpses remain buried under this enormous pile (Saab and Suleiman 2025). Unable to recover the bodies, families have written the names of those whom they have lost on the surface of bricks and concrete slabs, which now function as ephemeral tombstones. The rubble in Gaza is now a ghoulish amalgam of unexploded munition, concrete, steel, asbestos, glass, shrapnel, blood, bones and memories. A compound of decaying flesh, dead hopes and ordnance, it is both a graveyard and a minefield. The displaced build temporary shelters on it, and parasitic sandflies grow inside it. United Nations experts predict it will take fifteen years to clear the rubble in Gaza.

Rubble belongs to the 'after' of a singularity in time and space; its presence signals an event that has traumatically partitioned the flow of time into before and after. Rubble is what remains and stays after most of something has gone. Hence, it belongs to the order of the residual. We associate rubble with a temporary dimension, an in-between state, yet precisely because of – and not despite – its residual character, rubble is what stays. Rubble is a constant trait of aftermaths, and 'life in the rubble' is a condition far less temporary than we would like to think. This contribution is an invitation to stay with the rubble, to overcome our psychological reluctance to face the rubble – as remainder, incomplete, haunting, messy, dirty, traumatic – and delve into its materiality and plural lives. The material and semiotic afterlives of rubble encompass the zombie-like survival of debris (where the dead lie), the lives of those who died under the rubble, the lives of those who are living within the rubble, and the plural lives of the debris themselves.

To discuss this last point, it is important to highlight the tension between rubble and ruin. Drawing from Komska (2011), we understand 'ruin' (and its long intellectual history) as an abstract, widely symbolic signifier. Ruin is a fragment that always reminds us of a whole – a testimony, a monument to a whole that once was. By 'rubble' we designate the concrete physical debris associated with 'the end' – rubble is a noun that can be both singular and plural, as the individual rubble (as a piece, fragment of a pre-existent whole) does not exist. Rubble exists only as a multitude. In this chapter, we unpack the afterlives of ruination processes as they unfold in the immanence of the material and plural lives of rubble and call for geographical scholarship to engage with it.

Rubble, in this context, is a polyseme. The meanings and contestations it produces testify to the plural spatio-temporalities generated in its wake. The messiness of rubble offers itself as a resistance to organized singular interpretations and erasures. Unlike ruin, rubble is not a singular portal into a lost or imagined past or a desired future (see 'Does rubble matter?'). It continually renegotiates its relationship with time. The material and semantic existence of rubble as a multitude is mirrored in its temporal diffusion.

The plural lives of rubble and their spatio-temporal politics are what we are concerned with. In what follows, we propose a research agenda to study the geographies of rubble in the context of disasters. To demonstrate how we envisage future ethico-theoretical efforts, we review well-established research in human geography and cognate disciplines and highlight the need for research across disciplinary boundaries in three thematic areas where we see the potential for productive discussions, new questions and empirical analysis: political afterlives, ethnographies of rubble and rubble as counterarchives. We conclude by situating rubble research into ruination processes, agreeing with Stoler (2013: x), that 'centering on ruination shifts the emphasis from the optics of ruins to the ongoing nature of imperial process' and offer some reflections on how rubble can inform our relation to the past as well as our futures.

Does Rubble Matter?

The scant engagement with the materiality of rubble in disaster scholarship is hardly surprising when we consider the tension between rubble and ruin that has dominated Western approaches to what remains. We do not aim here at establishing a rigid dualism; rather, we discuss the articulations and implications of the spectrum spanning the distance from ruin to rubble to situate the invisibilities of rubble within specific intellectual histories and to highlight disciplinary blind spots (see Table 3.1) that are pertinent to disaster geographies.

Modern intellectual engagement and fascination with ruins are enmeshed in value creation, memory, history and nostalgia. As a part of a whole, ruins represent something that is missing, that we aspire to return to, to glorify, to remember and to preserve: ruins are 'mnemonic devices' (DeSilvey 2017). Even if we follow Simmel (1958), the power of ruin lies in its ability to become a new autonomous form, reversing the cosmic order from spirit to nature. As such, ruins are monuments, symbols, in a dialogical or even messianic relation with the past and the future (see Benjamin 1940). Ruins are worth being preserved and restored; they produce archival evidence of a past that we elect to remember. Elevating debris to the status of ruin is the essential gesture of value-based preservation through which empires erased the material traces of certain histories while producing history. There are countless examples of this in archaeological practice: the French expedition that excavated and published the Temple of Bel in Palmyra during the French Mandatory rule of Syria in 1929 decided to preserve the Greco-Roman past of the site while erasing – reducing to rubble – eight centuries of the site's history as a Muslim village (for an acute analysis of this 'imperial hierarchy of debris', see Katrib 2018). Ruins are protected by legislation and cultural norms and are subjected to processes of museification. From Romanticism's intellectual fascination with ruins to archaeology and contemporary memory studies, the analysis of ruins is fiercely anchored in the humanities. If we turn to rubble, we are dealing with the logical shadow of a ruin: a shapeless multitude as opposed to form – what is to be cleaned, cleared, forgotten and erased to leave space for reconstruction. Rubble is considered to

Table 3.1 A valued-based treatment of debris in the spectrum going from ruin to rubble

Ruin	**Rubble**
Part representing a whole (symbol) – fragment	A multitude, always in the plural
Thought to represent universal values that are here to stay (eternity) – associated with affects of identities, belonging, the sublime, history (official)	Thought to be ephemeral, associated with feelings of shame, loss, the uncanny, disorder, disposability
Memory (monument) – in a dialogical relation with the past and the future	To be forgotten – a present/presence to be overcome
Archival evidence, value	No recognized value, waste to be disposed of
Protected by legislation, cultural norms	Appropriated in an unregulated manner, neglected, production of (il)legality
Part of processes of museification, memory creation, revisionism	Part of ruination processes (ruin as a verb) (Stoler 2008)
Visible, chosen outcome to commemorate and glorify imperial processes	Imperial processes active in the present. Necropolitics: Whose lives are disposable?
Immobile, tethered to specific histories and forms of place-making, producing a politics of intangible value	Mobility and disposability engender a politics of volume and quantification

Source: Author created.

have no preservation value, in itself or symbolically, and is therefore thought of as disposable and ephemeral. The material mobility of rubble, unlike ruin, makes it spatio-temporally fluid, and as it is shifted from one place to another, it (re)produces and shapes new kinds of politics at different scales. Rubble tends to matter more in discussions around disposal, while the scholarship on ruin tends to converge around concerns of conservation.

Intrigued by the ruin/rubble polarization, we conducted a Scopus-based review of the literature on disasters published since 1990 that features the words 'rubble', 'debris' and 'ruin' in their abstracts. We left out the publication featuring the word 'rubble' only in their title as a metonymy to evoke 'disaster'. The value-based intellectual history that we tried to sketch above is unsurprisingly mirrored in disaster research. On the one hand, we have the subfield of disaster studies engaging with cultural heritage, which is concerned with the preservation of what international norms and organizations (e.g. UNESCO, OECD), as the institutional embodiment of a specific gaze, define as worthy ruins, including their preservation, restoration and archiving (Möller 2013; Schäfers 2016; Littlejohn 2021). On the other hand, rubble is equated to waste, and a majority of disaster studies spanning the disciplines of physical geography, engineering and environmental sciences are preoccupied with the disposal of rubble as waste management (see Brown et al. 2011 for a review).

What emerges from these publications is that the vast majority of articles concerned with debris discuss earthquakes (Faleschini et al. 2017; Sakai et al. 2019) and, to a much lesser degree, engage with floods (Gabrielli et al. 2018; Marchesini 2024). The terms remain undefined, and often 'rubble', 'ruin' and 'debris' are deployed interchangeably. None of the reviewed articles engages deeply with the materiality of rubble unless the publication is concerned with post-disaster waste management and recycling (Elias-Ozkan 2001; Pfatteicher 2010; Pollino et al. 2020). The classification of the typical debris streams for different types of disasters is usually drawn from the guidelines provided by FEMA (2007) (e.g. vegetative; construction and demolition; personal property/household items; hazardous waste; household hazardous waste; white goods; soil, mud and sand; vehicles and vessels; and putrescent (FEMA 2007; USEPA 2019)), followed by waste treatment options, recycling and waste-to-energy solutions. Indeed, we find that rubble is predominantly framed as either waste or ruin.

However, in this chapter, we argue that between the two ends of the spectrum, between ruin and waste, another semiotics of rubble is possible: one concerned with the permanence of rubble, its vibrant political and social lives, the value of its legal and illegal reuses, its affective histories and its visibilities.

Disaster geographies would benefit from a closer engagement with anthropology, where we find studies on the afterlives of rubble, their affective aftermaths and politics (Stoler 2008, 2013; Simpson 2013, 2020; Gordillo 2014; Gamberi and Calzana 2025), upon which we draw in what follows to sketch a research agenda organized around four broad areas in which rubble and ruination processes matter for disaster studies. These important contributions on 'the social life of ruins' (Dawdy 2010: 761), along with insight from the humanities, guide our attempt at sketching what disaster geographies will look like if we dare to consider the full extent of the spectrum of ruination. What if, instead of focusing solely on the opposite poles of heritage conservation and waste management, we let the rubble matter?

Rubble Visibilities: A Research Agenda for Disaster Geographies

The Political Afterlives of Rubble

Disaster geographies seem to be concerned with the two extremes of the spectrum, ruins as cultural heritage and rubble as waste, but what about the rubble that lies in between? Who sees that rubble? For whom does it matter? We argue that in rubble lies a plurality of identities and agencies that can help to explain the socio-economic and political aftermath of disasters. Rubble is not ephemeral; rather, it is a constant trait of aftermaths. The rubble produced in the wake of the San Juan earthquake of 1944 in Argentina became an 'indictment of the old political order' (Healy 2011: 4) and lubricated the pathway to power for Juan Perón, who used it to project his vision for a new Argentina, prompting Healy (2011: 8) to say, 'Peronism was born from the rubble; it is also a story of how Peronism was lived amid the enduring rubble, and despite it'.

In a similar vein, Edward Simpson (2013) has convincingly shown how the rubble of the 2001 Bhuj earthquake in Kachchh district in the western Indian state of Gujarat animated the 'political biography' of the earthquake's aftermath at different scales, from the trauma of rescuing and removing people from under the rubble, to the politics of removing rubble itself, and to rubble emerging as a locus of spectral haunting and a source of grief in the post-earthquake city. Similarly, in Mahesh Dattani's radio play, *Clearing the Rubble* (2013), the uneven outcome of a disaster is made visible through the lives of his Dalit Muslim protagonists in a village after the Bhuj earthquake, whose existence and survival matter little to state agencies. Rubble, for Dattani, is a metaphor for their lives before and after the earthquake, where the event serves as both a punctuation mark in and an illustration of their continued marginality in Gujarat.

While the politics of rubble in post-earthquake Bhuj has received limited attention (Mahadevia 2001; Simpson 2005, 2013, 2020; Simpson and Corbridge 2006; Simpson and de Alwis 2008; Simpson and Serafini 2019), we expand on this and subject it to our conceptual scrutiny by exploring the controversy surrounding a major case of corruption related to rubble removal in Bhuj – colloquially referred to as the 'malba scam' – to illustrate how rubble animates shifting spatio-temporalities in post-disaster landscapes.

Considered one of the deadliest earthquakes in India in the twentieth century, the 7.7 moment magnitude[1] Bhuj earthquake of 2001 produced approximately

11 million tonnes of rubble. The loss of almost 20,000 lives in the city, with a population of 136,000, was compounded as it took place on 26 January 2001 – a public holiday (Republic Day) – leaving more people indoors than otherwise. To speed up the removal of rubble, the Gujarat government enrolled private contractors – an early example of privatizing post-disaster response,[2] the stated aim of which was to address the shortfall of equipment and logistics of public agencies and voluntary organizations and to speed up recovery. However, a substantial volume of that 11 million tonnes of rubble was fraudulently reported to have been removed, and payment for 'completed' work was claimed through falsified paperwork. As news of the US$3 million scam emerged in the public domain, an intense government investigation was launched, and hundreds of people were arrested and imprisoned.

This intense punitive action, with legal cases being pursued against the accused even today, would shape the 'anti-corruption' image of the chief minister of Gujarat, Narendra Modi, who made the recovery of Bhuj a personal project to establish his claim as an able administrator. The public image created in Gujarat would also reify his national appeal in the 2014 elections, offering the opportunity for the incumbent Hindu nationalist Bharatiya Janata Party to position itself as an agent of 'good governance' and 'anti-corruption'. The rubble of 2001 was exploited to attach 'good governance' to the politics of Hindu nationalism, which deeply shaped national electoral politics in 2014. As such, we suggest that for the past two decades, the material and discursive politics of rubble in post-earthquake Bhuj has been an invisible yet key protagonist of the political life of Gujarat and India.

The malba scam illustrates that 'material objects are not passive surfaces on which social life is carried out' (Schäfers 2016). Rather, they embody, engender and animate social life. Indeed, rubble has a mobility that produces multi-scalar effects. For example, much of the rubble in Bhuj was illegally disposed of by contractors on Bhuj's ancient water harvesting structures, including water transporting channels, lake-beds and open wells. Indiscriminate dumping of rubble debilitated Bhuj's functioning water management system and groundwater aquifers, exacerbating the city's perpetual dependence on external sources like the Narmada river, located 500 kilometres away, and also on hyper-extractive technologies like reverse osmosis membrane-based water purifiers, which temporarily mask the crisis of Bhuj's exhausted aquifers. We identify similar observations on post-disaster rubble and its effect on local hydrology in Swamy's (2021) research on the Indian Ocean tsunami of 2004, where the violent shifting of rubble and silt by the waves clogged the mouths of rivers that were critical for mobility and livelihood in the fishing villages of the area. By drawing attention to this interaction of post-disaster rubble and hydrosocial processes, we highlight another missing aspect in disaster geography scholarship.

The city of Bhuj was founded in 1510 CE; hence the post-earthquake rubble of 2001 was composed of antique artefacts five centuries old – wooden doors with intricate carvings, door handles, door-knobs, stained glass panels, gables, corbels and tiles. The word 'malba' is a reference to waste or rubbish and not rubble. The word is borrowed from Hindi, as it doesn't exist in either Gujarati or Kachchhi,[3] indicating that the use of this word to denote useless material or rubbish is exogenous to the region and may have emerged from everyday usage of Hindi-speaking labourers and contractors. This unpacking of the term is critical to understanding how the semantic transformation of post-earthquake rubble into 'malba' worked towards devaluing the material residue of a disaster into waste that required disposal. As the semantic reconfiguration of rubble into malba devalued it of such historicity and heritage, converting these significant artefacts into worthless matter, the associated qualitative shift of heritage into quantifiable and disposable malba justified bureaucratic efficiency in disposing of rather than sorting through the rubble. However, contractors who recognized the value of antiquities and artefacts buried in the rubble facilitated their smuggling (Tyabji 2006).

We return to Clearing the Rubble (Dattani 2013) to illustrate that the construction of 'malba' in post-earthquake Bhuj intersects with questions of marginality. As part of his doctoral research, one of us conducted interviews in April 2018 with senior residents of Bhuj who survived the 2001 earthquake. In these interviews, there was repeated reference to 'outsiders' who took advantage of the disaster to steal from unoccupied homes. However, a closer examination of ground reports from the archives of Kachchh Mitra newspaper (Kachchh Mitra 2001) reveals that they were not just 'outsiders' geographically but also demographically. By placing archival reports and interview data together, it emerged that 'outsiders' did not exclusively indicate people from outside Bhuj or Kachchh but daily wage labourers from Ahmedabad and other city centres in Gujarat who had arrived in Bhuj to earn a living through the removal of rubble. Many of these workers were Dalits or from lower-caste groups. The thought of their unsanctioned access to upper-caste homes – which would be unthinkable if those premises were occupied – unsettled residents. In the interviews, respondents mentioned that people preferred to secure valuables such as gold and silver ornaments and cash in their homes, especially those from the Jain community, and the possibility of finding those items in abandoned premises could have led to incidents of theft. In-depth interviews with journalists who reported on the post-earthquake city revealed that, though these incidents were not widespread and, as per official records, there were no any reports of theft or arrests, the intense indignation they produced was fuelled by perceptions of transgression caste and religious barriers.

For higher-caste people, the ability of lower-caste Dalits to enter their homes was as unsettling as the material ruination of their homes and was viewed by them as a humiliating and unforeseeable byproduct of their post-earthquake misfortunes. Rubble was not just composed of abiotic residue; it also contained corporeal remains of the dead, and that admixture, even though reframed as malba, imbued it with both affection and disgust. The latter enabled the transgression of caste-based socio-spatial barriers of untouchability, as the labour of moving decaying corpses was forced on Dalits. In his review of studies on post-disaster relief and rehabilitation in India, Parthasarathy (2015) argues that caste-based discrimination plays a significant role in the way labour is apportioned. Referring to the Mumbai floods of 2005, he states:

> Thus, while relief workers after the floods were officially in the field and had access to bulldozers, cranes, and other equipment, the actual task of sifting through, clearing, and cleaning up the accumulated garbage, rubble, debris, and dead bodies (of humans and animals) fell upon municipal conservancy workers, ragpickers, and other unemployed low-caste men (women were rarely involve in the cleaning up, except at household levels).
>
> – Parthasarathy (2015: 565)

In this context, the framing of rubble as 'malba' acquires salience, as it intersects with questions of marginality in post-disaster landscapes. As we notice, people and rubble are mutually entangled; hence we need to approach the study of rubble geographies relationally: their connection is to be understood as a political and historical singularity (Kovač and Ramella 2023). For instance, when studying the destruction and afterlife of rubble in the Gran Chaco region of Argentina, Gastón Gordillo (2014: 81) argues that destruction disintegrates not only physical traces of the rubble but also 'the conditions of sociality that define a particular spatial node'. The visibility of such spatial nodes is what disaster geographies should ceaselessly seek in the rubble.

The malba scam is an empirical illustration of the political afterlife of rubble, where it emerges not as a passive substrate but as vibrant matter animating post-disaster politics. As we show, it becomes an active ingredient in influencing urban governance, as concerns of caste-based inequality and nationalist politics coalesce around it, allowing it to exert influence across scales.

While drawing attention to the multi-scalar aspect of the politics of rubble, we also recognize that rubble requires closer and more intimate attention, which is usually obliterated when approaching its materiality with prefabricated and preoccupied hierarchies of values. In the next section, we delve deeper into how ethnographic research on rubble could enrich disaster geographies.

'The Legal Production of Illegality': Ethnographies of Rubble

In disaster geography, the production of rubble is viewed as episodic. We argue that rubble is also produced through everyday acts of violence and ruination. Settler-colonial violence reduces Palestinian homes to rubble, and in a similar vein, Muslims in India are increasingly being subjected to 'bulldozer justice', where homes are unconstitutionally razed if any family member is accused of committing a crime. There are less obvious ways in which ruination processes produce rubble in connection with disasters, which intersect various temporalities and function as yet another reminder of how inadequate the linearity of the mainstream 'disaster cycle' paradigm can be.

As articulated in the previous sections, the conceptual mattering of rubble unveils the multidimensional temporalities that shape the political act of ruination in disaster aftermaths. Such ruination is also located at the core of urban planning and animates the decision to turn homes, shops and entire neighbourhoods into rubble through the necropolitical weaponization of illegality. To bring about regeneration, requalification or reconstruction in the aftermath of a disaster, as well as preventative disaster reduction measures, requires listening to the stories in the rubble and uncovering whose lives were deemed disposable and whose existences were turned into rubble, destroyed and/or rebuilt with them. As argued by Brunotti (2021) in the context of Zanzibar, only 'an ethnography of rubble will make apparent spatial illegalization, i.e. the legal mechanism through which planners produce illegality' (Roy 2019, as cited in Brunotti 2021: 77). An analytical focus on rubble foregrounds the violence of settler colonialism, sectarian politics and big-D development-funded disaster risk mitigation processes that produce illegality as a means to evict urban villages with long and complex histories, where land tenure is informal and evasive, from the kampungs in Jakarta (Tilley et al. 2017) to the *katchi abadis* in Islamabad/Rawalpindi (Mustafa et al. 2015). Removal of slums as 'encroachments' in Bangalore around drainage areas is similarly projected as necessary for developing urban 'flood resilience', legitimizing the targeted eviction of Dalit populations (Ranganathan 2022). A similar process occurred in post-earthquake Bhuj, where rubble was dumped in Prag Sar, a lake that functioned as a rainwater harvesting and groundwater recharge structure – key processes for disaster risk mitigation and climate change adaptation. The dumped rubble led to the erasure of the lake, and with it, Bhuj became more vulnerable to local hazards. The transformation of Prag Sar from an open access common to an enclosed government property was enabled through the disposal of rubble.

Labelled as illegal 'slums', urban villages are often reduced to rubble to make space for donor-driven technofixes (e.g. the Dutch polder system in Jakarta or the inappropriate Japanese top-down flood early warning system in Pakistan), which have inconsistent success across contexts. These ruination processes are often accompanied by poorly designed relocation programmes, which often lead to villagers refusing to leave and choosing to remain in place. In such circumstances, villagers must rebuild their existence within the rubble of their former lives, exposing them to health hazards caused by the dust and poisons, dispossession, loss of livelihoods, grief and trauma.

Similarly, what has been popularized under the rubric of the 'disaster capitalism complex' (Klein 2007) forces us to consider also the ex-post-disaster phase. Well-studied examples include the post-2004 tsunami eviction of coastal Sri Lankan fisherfolk to make room for high-rise hotels for luxury tourism and the gentrification of New Orleans' floodplains. The rubble exposes processes of ruination and imperial histories that are otherwise hidden under seemingly unobjectionable good values such as fighting 'environmental degradation', contributing to 'risk mitigation', 'development' and, of course, 'resilience'. By conducting ethnographies of rubble, disaster geographies can uncover the forced or enforced wreckage that can span decades and rest undetected beneath disaster risk reduction jargon and 'build back better' slogans.

Ethnographic investigations of rubble reveal hidden histories of inequality, vulnerability and neglect. They also illustrate how it is part of a long-term cycle of ruination. An interesting example comes from the periphery of Mexico City, where the government illegally dumped debris into the lake-bed swamp following the 1985 earthquake. The filled-in lake-bed was then urbanized as Colonia del Mar. In 2017, another major earthquake caused structural collapses and damage along the lake-bed edge of Mexico City, where Colonia del Mar is located (Moreno-Carranco and McElvain 2020). This *fil rouge* between the rubble of two major earthquake events exposes long-term ruination processes connecting two aftermaths and indicates how approaches to relocation and rebuilding are profoundly informed by historical processes that socially produce risk in marginalized areas.

Ruination processes critically pertain to the decision to remember and archive or destroy and annihilate, and it is to this dimension that we turn in the next section.

Connectors: Rubble as Counterarchives

Disaster research has increasingly recognized the importance of moving beyond the 'exceptionality' paradigm and focusing on the everyday, the thick, uneventful temporalities in which vulnerabilities and inequalities are built, are stratified and ultimately catalyse the event of a disaster (Kelman 2020). Less attention has been paid to ruination processes and to how they crystallize around debris in a variety of ways that exceed the polarized attention to either their disposal (as waste) or their memorialization (as ruins).

We argued that within this spectrum, ethnographic work on rubble is a powerful method to expose ruination processes and the relational forces that they entail. As discussed in the previous sections, rubble constitutes an untapped archive of counter-histories and a repository for alternative ways of relating to the past and creating our futures. Rubble is a spatial rearrangement that is also a laceration of the ordinary flow of time. Such disruption of habitus tells us invaluable stories of how social rules, norms and bonds are formed, as well as how they can be changed or subverted. This means that from the rubble, we can learn social precarity and structural violence as an aetiology of disasters, as well as benefit from its generative power that can lead to both dystopic and utopic futures (Dawdy 2010).

We have been taught the power of the archive (Derrida, Foucault), the political power descending from control over the archive. As history glorifies its ruins through conservation and various types of digitized preservation, the materiality of rubble speaks to another type of archive, one that is not afraid to let rubble's vital matter vibrate (Bennett 2010). We cannot engage here in a discussion on agency, but, with Leen Katrib (2018), we dare pose the question: How can rubble function as an archive? In particular, we pose this question to scholars concerned with disasters in an attempt to give visibility to the relational character of rubble and to its plurality and complexity. Rubble is a site of resistance, first of all because it stays, despite the degree of destruction and violence that has been unleashed to engender it. Furthermore, rubble, as a multitude, resists linear classification based on order and categorization and points instead to the complexity of multiple, overlapping and contested gazes and meanings. Rubble is a plurality of views and can be seen at the same time as a trace of what is lost, desolation, violent displacement, reclamation of space, available material to profit from, a site of aesthetic ponderations, political propaganda or colonial ruination (DeSilvey and Edensor 2012; Le Mentec 2025: 46).

Such plurality is what makes rubble alive, vibrant matter that does not ossify memory into the vestige of a (single) past to remember but rather points at embodied contestations. Rubble is well understood as a complex assemblage: in its presence 'past hauntings, materiality, socio-political framework, subjective interpretations, and perceptions are intricately intertwined' (Gamberi and Calzana 2025: 7). Ethnographies of rubble in disaster aftermaths can then collect traces of history, memory, identities and affects that are situated and can serve as a record of collective experiences and traumas. However, it would be a mistake to stop at negative effects. Rubble offers prosperity and economic opportunities for some. Rubble can be a cultural trace of trauma as well as an archive of survival and resistance. The stubborn counter-histories that people rebuild from the broken remnants of their former lives. In terms of material culture, excavating rubble might reveal discarded objects, personal belongings or artefacts that speak to the daily life of people before the destruction (Sou and Webber 2019). They can be subjected to cultural, linguistic and political interpretation but also denote the wound and the personal affects of touching detail that establish a direct relationship with the object or person within it.

An important and neglected question for disaster studies is how we preserve the memory of a disaster, how disaster memorials are produced and which archives they mobilize. We see the potential here for dialogue and cross-fertilization with memory studies to further our understanding of how rubble is vibrant and alive in the contestation of memory. After the 2008 Wenchuan earthquake in China, the Chinese government initiated a process of 'heritagization', which is the transformation of disaster-affected areas and cultural practices into heritage sites to promote economic recovery and cultural preservation, with the double intent of preserving the memory of the disaster and promoting tourism to the affected area. In her anthropological studies in Beichuan and Wenchuan, Le Mentec shows the disappointing outcome of post-disaster heritage programmes, which neglected locals' points of view, practices and affects, while promoting dark tourism through the commodification of the traditions of the particularly affected Qiang minority, without delivering the promised economic 'development' (Le Mentec and Zhang 2017; Le Mentec 2025). Le Mentec shows, through powerful ethnographic vignettes, the lives and meanings of rubble that are obliterated by top-down post-disaster heritage concerns and plans, pointing once again to the urgent need to engage with ethnographic work on rubble.

The Future of Rubble

Rubble needs to move from silence to salience in disaster geographies. As we have illustrated, rubble has a political afterlife, and it can reproduce risk and animate questions of development and urban planning. Rubble has to do with counter-histories and with decisions to delete certain pasts and memorialize others. If ruins induce 'life' into immutable structures, then rubble, through its fluidity, mobility, plasticity and durability, requires critical engagement.

Rubble is also a powerful future-making device that acts in ruination processes, shaping our present and inhabiting our imaginaries. Kovač and Ramella (2023: 142) argue that 'futures emerge from ruins and rubble of undelivered and uncertain promises'. Building on this idea, we suggest that a focus on rubble would work to advance debates on 'future-making' in disaster geographies. For example, in Gaza, the rubble has been repurposed to legitimize ethnic cleansing, with US president Donald Trump arguing that 'relocating' Palestinians in Gaza is a 'solution' because they would otherwise be condemned to live in precarity in the ruins and rubble of collapsing built infrastructure. Following the erasure of Palestinians from their land, Gaza is offered as a site for waterfront real estate development or, as Trump himself put it, a 'Riviera of the Middle East' (Al Jazeera 2025). In his blog for the Atlantic Council, a think-tank based in Washington, DC, Ahmed Fouad Alkhatib (Alantic Council 2025) elaborates on the 'short-term' 'future-making' of rubble: 'Gaza's rubble should be recycled and used to create an artificial peninsula off its central coast. That peninsula could host an airfield and seaport that could serve as a non-Israeli and non-Egyptian lifeline as part of the Strip's recovery and rebuilding'. In these ways, the raw material of rubble is taken advantage of to justify the genocidal reworking and new imperialist imaginaries. In contrast, Palestinians returning to North Gaza are using the rubble to remake their homes. In so doing, they engage in a process of reclaiming their material and emotional pasts and reject the 'future-making' of their oppressors. As such, the case of Gaza demonstrates how rubble can signal simultaneous acts of rupture and conjoining, a material-semiotic connection between potential future lives and past lives.

Acknowledgements

This research was supported through the 2023/24 HEQR Seed Funding at Bath Spa University. We are also thankful to Karamveer Jadeja for supporting us with archival research in Bhuj and to Rebecca Speich for her contribution to the classification of the literature we reviewed for this chapter.

Notes

1 Moment magnitude scale.

2 The Gujarat government would later go on to privatize post-disaster rehabilitation in rural and urban Kachchh through a 50:50 public–private partnership (Mahadevia 2001).

3 The closest term we could locate is ભંગાર (bhangar), meaning broken pieces of wood and metal.

References

Al Jazeera (2025), 'Trump Says US Will "Take over" and "Own" Gaza in Redevelopment Plan'. Available at: https://www.aljazeera.com/news/2025/2/5/trump-says-us-will-take-over-and-own-gaza-in-redevelopment-plan (accessed 7 November 2025).

Atlantic Council (2025), 'Ahmed F. Alkhatib'. Available at: https://www.atlanticcouncil.org/expert/ahmed-alkhatib/ (accessed 7 November 2025).

Benjamin, W. (1940), 'Über den Begriff der Geschichte', in *Walter Benjamin zum Gedächtnis*, 1942, Die Neue Rundschau.

Bennett, J. (2010), *Vibrant Matter: Kachchh A Political Ecology of Things*, Duke University Press.

Brown, C., M. Milke and E. Seville (2011), 'Disaster Waste Management: A Review Article', *Waste Management* 31 (6): 1085–98. https://doi.org/10.1016/j.wasman.2011.01.027.

Brunotti, I. (2021), 'Kifusi: Towards an Ethnography of Rubble', *Stichproben: Vienna Journal of African Studies* 21 (41): 63–82. https://doi.org/10.25365/phaidra.310_04.

Dattani, M. (2013), *Clearing the Rubble*, Penguin.

Dawdy, S. (2010), 'Clockpunk Anthropology and the Ruins of Modernity', *Current Anthropology* 51 (6): 761–93.

DeSilvey, C. (2017), *Curated Decay: Heritage Beyond Saving*, University of Minnesota Press.

DeSilvey, C. and T. Edensor (2012), 'Reckoning with Ruins', *Progress in Human Geography* 37 (4): 465–85.

Elias-Ozkan, S. T. (2001), 'Recycling Rubble into Aggregates: A Model for Local Governments', *Habitat International* 25 (4): 493–502. https://doi.org/10.1016/S0197-3975(01)00020-0.

Faleschini, F., Z. Mariano Angelo, L. Hofer, P. Zampieri and C. Pellegrino (2017), 'Sustainable Management of Demolition Waste in Post-Quake Recovery Processes: The Italian Experience', *International Journal of Disaster Risk Reduction* 24:172–82.

FEMA (2007), 'Public Assistance Debris Management Guide', July. Available at: https://www.fema.gov/sites/default/files/2020-07/fema_325_public-assistance-debris-mgmt-plan_Guide_6-1-2007.pdf (accessed 7 November 2025).

Gabrielli, F., A. Amato, S. Balducci, L. Magi Galluzzi and F. Beolchini (2018), 'Disaster Waste Management in Italy: Analysis of Recent Case Studies', *Waste Management* 71:542–55. https://doi.org/10.1016/j.wasman.2017.10.012.

Gamberi, V. and C. Calzana (eds) (2025), *Haunting Ruins: Ethnographies of Ruination and Decay*, Berghahn.

Gordillo, G. (2014), *Rubble: The Afterlife of Destruction*, Duke University Press.

Healey, M. A. (2011), *The Ruins of a New Argentina: Peronism and the Remaking of San Juan After the 1944 Earthquake*, Duke University Press.

Kachchh Mitra (2001), *Special Earthquake Memorial Archive Issue*, Bhuj.

Katrib, L. (2018), 'On Archiving Rubble', *Future Anterior: Journal of Historic Preservation, History, Theory, and Criticism* 15 (2): 33–45. https://doi.org/10.5749/futuante.15.2.0033.

Kelman, I. (2020), *Disaster by Choice*, Oxford University Press.

Klein, N. (2007), *The Shock Doctrine: The Rise of Disaster Capitalism*, Picador.

Komska, Y. (2011), 'Ruins of the Cold War', *New German Critique* 38 (1): 155–80.

Kovač, U. and A. L. Ramella (2023), 'From Ruins and Rubble: Promised and Suspended Futures in Kenya (and Beyond)', *Journal of Eastern African Studies* 17 (1–2): 141–64. https://doi.org/10.1080/17531055.2023.2245263.

Le Mentec, K. (2025), 'From Desire to Avoidance: Contrasting Interactions with the Rubble of Familiar Spaces Affected by an Earthquake and a Planned Flood (China)', in V. Gamberi and C. Calzana (eds), *Haunting Ruins: Ethnographies of Ruination and Decay*, 45–67, Berghahn.

Le Mentec, K. and Q. Zhang (2017), 'Heritagization of Disaster Ruins and Ethnic Culture in China: Recovery Plans After the 2008 Wenchuan Earthquake', *China Information* 31 (3): 349–70. https://doi.org/10.1177/0920203X17736508.

Littlejohn, A. (2021), 'Ruins for the Future: Critical Allegory and Disaster Governance in Post-Tsunami Japan', *American Ethnologist* 48 (1): 7–21. https://doi.org/10.1111/amet.13006.

Mahadevia, D. (2001), 'Privatising Earthquake Rehabilitation', *Economic and Political Weekly* 36 (39): 3670–73.

Marchesini, G. (2024), 'How Does the Waste Management System Reorganised in Response to a Disaster? A Case of the Roya Valley (France) After Storm Alex', *International Journal of Disaster Risk Reduction* 116:105052. https://doi.org/10.1016/j.ijdrr.2024.105052.

Möller, M. (2013), *Rubble, Ruins and Romanticism: Visual Style, Narration and Identity in German Post-War Cinema*, transcript Verlag.

Moreno-Carranco, M. and B. McElvain (2020), 'Life Above, Rubble Below: A Case of Historically Produced Risk & Perception in Mexico City', *Ardeth* 6:153–69. https://doi.org/10.17454/ARDETH06.12.

Mustafa, D., G. Gioli, S. Qazi, R. Waraich, A. Rehman and R. Zahoor (2015), 'Gendering Flood Early Warning Systems: The Case of Pakistan', *Environmental Hazards* 14 (4): 312–28. https://doi.org/10.1080/17477891.2015.1075859.

Parthasarathy, D. (2015), 'Informality, Resilience, and the Political Implications of Disaster Governance', *Pacific Affairs* 88 (3): 551–75. https://doi.org/10.5509/2015883551.

Pfatteicher, S. K. A. (2010), *Lessons Amid the Rubble: An Introduction to Post-Disaster Engineering and Ethics*, Johns Hopkins University Press.

Phipps, A. (2025), 'Rubble', Instagram. Available at: https://www.instagram.com/alisonphipps/p/DFXI2aYNutz/ (accessed 9 April 2025).

Pollino, M., S. Cappucci, L. Giordano, D. Iantosca, L. De Cecco, D. Bersan, V. Rosato and F. Borfecchia (2020), 'Assessing Earthquake-Induced Urban Rubble by Means of Multiplatform Remotely Sensed Data', *ISPRS International Journal of Geo-Information* 9 (4): 262. https://doi.org/10.3390/ijgi9040262.

Ranganathan, M. (2022), 'Caste, Racialization, and the Making of Environmental Unfreedoms in Urban India', in J. F. Cháirez-Garza, M. D. Gergan, M. Ranganathan and P. Vasudevan (eds), *Rethinking Difference in India through Racialization*, 43–63, Routledge. https://doi.org/10.4324/9781003319726-3.

Saab, S. H. and R. Suleiman (2025), '50,000 Palestinian's Have Been Killed in Gaza', *Haaretz*. Available at: https://www.haaretz.com/israel-news/2025-03-24/ty-article-static/.premium/50-000-palestinians-have-been-killed-in-gaza-this-is-how-it-happened-day-by-day/00000195-5706-d544-a795-df9eacaa0002 (accessed 9 April 2025).

Sacco, J. (2024), 'War on Gaza', *Comics Journal*, 25 June.

Sakai S., R. Poudel, M. Asari and T. Kirikawa (2019), 'Disaster Waste Management After the 2016 Kumamoto Earthquake: A Mini-Review of Earthquake Waste Management and the Kumamoto Experience', *Waste Management & Research* 37 (3): 247–60. https://doi.org/10.1177/0734242X18815948.

Schäfers, M. (2016), 'Ruined Futures: Managing Instability in Post-Earthquake Van (Turkey)', *Social Anthropology* 24 (2): 228–42. https://doi.org/10.1111/1469-8676.12274.

Simmel, G. (1958), 'Two Essays', *Hudson Review* 11 (3): 371–85.

Simpson, E. (2005), 'The "Gujarat" Earthquake and the Political Economy of Nostalgia', *Contributions to Indian Sociology* 39 (2): 219–49. https://doi.org/10.1177/006996670503900202.

Simpson, E. (2013), *The Political Biography of an Earthquake*, Oxford University Press.

Simpson, E. (2020), 'Forgetfulness Without Memory: Reconstruction, Landscape, and the Politics of the Everyday in Post-Earthquake Gujarat, India', *Journal of the Royal Anthropological Institute* 26 (4): 786–804. https://doi.org/10.1111/1467-9655.13416.

Simpson, E. and S. Corbridge (2006), 'The Geography of Things That May Become Memories: The 2001 Earthquake in Kachchh-Gujarat and the Politics of Rehabilitation in the Prememorial Era', *Annals of the Association of American Geographers* 96 (3): 566–85. https://doi.org/10.1111/j.1467-8306.2006.00706.x.

Simpson, E. and M. de Alwis (2008), *Politicising Grief and Memory in Post-Disaster Gujarat and Sri Lanka, Non-Governmental Public Action Programme Working Paper Series No. 33*, London School of Economics and Political Science.

Simpson, E. and M. Serafini (2019), 'Earthquake Citizens: Disaster and Aftermath Politics in India and Nepal', in M. Holbraad, B. Kapferer and J. F. Sauma (eds), *Ruptures: Anthropologies of Discontinuity in Times of Turmoil*, 193–217, UCL Press. Available at: http://www.jstor.org/stable/j.ctvf3w1v4.15 (accessed 12 September 2025).

Sou, G. and R. Webber (2019), 'Disruption and Recovery of Intangible Resources During Environmental Crises: Longitudinal Research on "Home" in Post-Disaster Puerto Rico', *Geoforum* 106:182–92.

Stoler, A. L. (2008), 'Imperial Debris: Reflections on Ruins and Ruination', *Cultural Anthropology* 23 (2): 191–219. https://doi.org/10.1111/j.1548-1360.2008.00007.x.

Stoler, A. L. (2013), *Imperial Debris: On Ruins and Ruination*, Duke University Press. https://doi.org/10.1215/9780822395850.

Swamy, R. (2021), *Building Back Better in India: Development, NGOs, and Artisanal Fishers After the 2004 Tsunami*, University of Alabama Press.

Tilley, L., J. Elias and L. Rethel (2017), 'Undoing Ruination in Jakarta: The Gendered Remaking of Life on a Wasted Landscape', *International Feminist Journal of Politics* 19 (4): 522–9. https://doi.org/10.1080/14616742.2017.1364907.

Tyabji, A. (2006), *Bhuj: Art, Architecture, History*, Mapin.

US Environmental Protection Agency (USEPA) (2019), 'Guidance About Planning for Natural Disaster Debris'. Available at: https://www.epa.gov/homeland-security-waste/guidance-about-planning-natural-disaster-debris (accessed 12 September 2025).

4

Reassembling Disaster Geographies

Placing the Material and Discursive

Peter McGowran

As the introduction to this edited collection has indicated, the discipline of geography has had a significant influence on the study of hazards and disasters. Nonetheless, several ambiguities and misunderstandings remain regarding what geographical theories mean for the study of disasters. At a time when interdisciplinary work is becoming more common, it is useful to reflect on what geographers can bring to the interdisciplinary table. In what follows, I will reflect on this by thinking with the broadly defined concept of 'assemblage', a key conceptual device and theoretical approach in contemporary critical geography. Assemblage-thinking has been used by geographers not only to frame research on 'what we call disasters' (Gaillard 2021) but also to critically reflect on the knowledge systems which underpin this research (Donovan 2020). I advance these ideas here by arguing that assemblage-thinking provides a useful vocabulary and way of thinking when reflecting on what anglophone geography, as a discipline and knowledge system, brings to the study of disasters and what this means for research on disasters and disaster risk. After first briefly situating my interpretation of assemblage-thinking with respect to the geographical literature, I break down the potential contributions of assemblage theory into three thematic sections on natures, temporalities and power–knowledge relationships in disaster management research and practice. In the end, I argue that assemblage geographies can lead researchers to more nuanced inquiries into the nature of disasters and what can or cannot be done to address them.

The Nature of Assemblages

It was in geographical debates about 'nature' in the 1990s and early 2000s that the concept of assemblage began to emerge in the geographical literature. These debates arose from the understanding that the idea of an 'external nature' was fundamentally rooted in the geographical imaginations which underpinned European colonialism (Castree and Braun 2001; Castree 2005; Smith 2010). Critical geographers had thus largely agreed that attempts to separate what is natural and what is not are futile at best and problematic at worst. In response, they sought out and debated the merits of alternative vocabularies and theoretical frameworks that could support their analysis of the complex interplay of processes which define space and place. Those coming from a Marxist perspective drew on the work of geographers like Neil Smith and Phil O'Keefe (1980) to talk about first and second natures, political ecologies (Blaikie and Brookfield 1987) and other such terms to argue that what has been understood as 'nature' in the past is, more often than not, tied into capitalist means of production in both material and discursive ways (Swyngedouw and Heynen 2003; Heynen et al. 2006). Such approaches were influential and indeed underpinned some of the most well-established theories of disaster causation in geography and beyond: the pressure and release model (Wisner et al. 2003) and O'Keefe et al.'s (1976) foundational argument that there is no such thing as a natural disaster. Nonetheless, such theoretical approaches attracted criticism from many geographers (e.g. Marston et al. 2005), and often those working in postcolonial contexts (McFarlane 2011). They argued that the Marxist foundations of some of these theoretical approaches meant that the influence of capitalist forces tended to be artificially separated from, or emphasized over, other influential processes that escape the logics of a solely Marxist explanation (Grove 2009).

Simultaneously, geographers also looked elsewhere for conceptual frameworks to make sense of the world around them. Francophone post-structural philosophy from the mid-twentieth century became a key resource in this regard, particularly following several important translations in the 1980s and 1990s. From here, geographers began to apply the ideas of Foucault, Deleuze and Derrida in their work on space, place and power. Additionally, contemporary interpretations and theoretical frameworks grounded in the work of those earlier French philosophers became particularly influential. One such example is actor-network theory (ANT), most often associated with Bruno Latour (2005) but also significantly influenced by several others (Law 1992; Mol and Law 1994; Murdoch 1997). Another, and arguably more influential theory in recent times, is DeLanda's (2016) assemblage theory, which is grounded somewhat more explicitly in the writings of Gilles Deleuze and Félix Guattari than ANT.

The adoption of these theoretical frameworks led geographers to write about things previously described as '(not) natural' in terms of 'non-human actors', 'more-than-human processes' and hybridized assemblages and networks of material and discursive processes (Whatmore 2002; Braun 2005; Robbins and Marks 2010). Indeed, ANT and assemblage theory were often conflated because they tended to underpin analyses that blurred, problematized and ultimately did away with the nature–society boundary. This conflation has been comprehensively critiqued by geographers since (e.g. Müller and Schurr 2016). As a result of these critical appraisals, assemblage theory emerged as the more dominant of the two theories in geography, perhaps on the basis that it is more explicitly grounded in Deleuzian philosophy – an exercise centred on highlighting and challenging domination and the over-concentration of power (Buchanan 2000). In that sense, it stood up more robustly than ANT to some recurrent critiques – namely, that both theories depoliticize and obscure the political and economic drivers of social and environmental injustice because of their unconventional vocabularies and tendency to distribute causation and blame across various processes and actors rather than more squarely pinning blame and causation on that which is political and economic (Lave 2015). As a result of these critiques, and particularly in the discipline of geography, contemporary interpretations of assemblage theory tend to actively incorporate and focus on political and economic processes and have sought to drop the capitalized 'assemblage theory' title in favour of 'assemblage-thinking' (McFarlane and Anderson 2011). These moves are taken with a view to addressing its perceived habit of depoliticization and the idea that it is a theory which provides a standardized, uncontested and homogeneous means of viewing the world: the latter being antithetical to any concept that claims to be 'Deleuzian' (Buchanan 2000).

All this being the case, and before advancing further with the chapter, it is worth briefly reflecting on what 'an assemblage' is, at least for geographers. In his book *Assemblage Theory*, which outlines his interpretation of the concept, DeLanda (2016: 1) bases his interpretation on (one of) Deleuze's translated reflections on what an assemblage is:

> What is an assemblage? It is a multiplicity which is made up of many heterogeneous terms and which establishes liaisons, relations between them, across ages, sexes, and reigns – different natures. Thus, the assemblage's only unity is that of a co-functioning: it is a symbiosis, a 'sympathy'.
>
> – Deleuze and Parnet (1987: 69)

It is the idea of an assemblage – as gathered, co-functioning, and related yet discrete processes – that defines geography's engagement with the concept (Anderson and McFarlane 2011). These processes can be inclusive of 'different natures' while also explicitly concerned with what is social, with all processes spanning the material and immaterial. Another crucial yet perhaps underappreciated component of this interpretation is that assemblages are processual. *Agencement*, the original French term for 'assemblage', captures this idea more easily. Nonetheless, a processual understanding of assemblages sees them not as fixed 'things' in space and time. Rather, they are always understood as unfolding constellations of interrelated but discrete processes. Thus, while the behaviour of assemblages is always subject to change, their unfolding can also be understood as an ongoing repetition, giving assemblages the impression of being fixed, stable and identifiable objects (Legg 2011). Following Deleuze's explanation given earlier, we might further speculate that he viewed processes often associated with hazards and disasters as amenable to an assemblage lens:

> It is never filiations which are important, but alliances, alloys; these are not successions, lines of descent, but contagions, epidemics, the wind.
>
> – Deleuze and Parnet (1987: 69)

With this understanding of assemblages in mind, the remainder of the chapter consists of sections that argue, respectively, that an assemblage approach can help scholars provide more nuanced accounts of how natures, temporalities and power relations shape disasters and people's experiences of them.

Reassembling Disaster Natures

In this section, I will suggest that assemblage-thinking, as it is understood in the geographical literature, can help take analyses of disasters beyond simplistic conclusions about their 'nature' towards more grounded and incisive conclusions about their causation.

That there is no such thing as a natural disaster is something that geographers and other scholars under the broad umbrella of disaster studies can agree on. Rather, it is widely understood that vulnerability to hazards is constructed, created or produced by political, economic and cultural processes, or that it is 'socially constructed' (Wisner et al. 2003; Kelman et al. 2016). Going further, Kelman (2018) argues that 'hazards' only become such once they pose a threat to society and that the hazards themselves may be the outcome of anthropogenic interventions (e.g. river channel manipulation, road construction). Hazards, too, he argues, might be understood as at least partly socially constructed. Taking all this into account, it has been suggested that disasters themselves are best understood as socially constructed (Oliver-Smith et al. 2017; Alcántara-Ayala et al. 2021). By extension, it is argued that addressing socially constructed vulnerability is seen as the primary means through which the risk of socially constructed disasters can be reduced (Fraser et al. 2014). This literature and logical explanation lie at the heart of disaster risk reduction as an official policy goal and are clearly embedded in the pages of the United Nations Sendai Framework for Disaster Risk Reduction (2015–30) (UNDRR 2015).

While attentive to the historical construction of social vulnerability, the argument that disasters are socially constructed has lost sight of some of the more nuanced and historically grounded analyses of 'nature' as an idea, realm or thing that gave rise to what is sometimes known as the 'vulnerability paradigm' in the first place (e.g. Smith and O'Keefe 1980; Watts 1983). Nonetheless, at a practical level, those firmly within a geographical tradition and those coming from the broadly defined field of 'disaster studies' both agree that disasters are, by and large, social phenomena (Donovan 2017). Geographers, though, may feel uneasy about the concept of 'social construction' being applied as a means to arrive at these ends, particularly when compared with the idea that disasters are not 'natural'. At a pragmatic level, this is because geographers feel such a framing raises the aforementioned debates which argue that setting up a dualism between things that are either 'natural' or 'socially constructed' is a dead end: it is well established that nothing ever exists entirely within one or the other realm. The term 'construction' is also vague and invites misunderstandings, particularly when trying to communicate such arguments to a non-academic audience. Demeritt (2002) elaborated on these issues at some length and depth in an influential piece some twenty years ago. Rather than recount the theory-dense arguments elaborated there, I will reflect on the limitations of a 'disasters as socially constructed' lens in relation to landslide disasters across the Himalayas.

The Hindu–Kush–Himalaya range stretches from central Asia eastwards to the north-east of India, China and Myanmar. The range is formed by the ongoing collision of the Indian and Eurasian tectonic plates, which causes the Himalayas to rise every year (Soja and Starkel 2007). Across much of the Himalayan range, these geological processes render the rocks and soils loose and unstable while creating very steep slopes. The collision of the two plates also generates pressure, which is released periodically, causing some of the most powerful earthquakes ever recorded. Much of the central belt of the Himalayan range, from western India through Nepal, Bhutan and north-east India, also receives monsoon rainfall (Starkel and Sarkar 2014). This combination of processes alone, spanning the geological, geomorphological, seismological and hydrological, is enough to generate mass movements or slope failures: they are part and parcel of Himalayan geomorphology (Starkel 1979). Drawing from Demeritt's (2002) working definition of how to define natural and/or not natural, it can be said that (some) landslides in the Himalayas have occurred, and will continue to occur, due to processes that are not socially contingent (i.e. they may be understood by some as 'natural'). These processes are certainly not 'socially constructed'.

While not socially contingent, these processes can only ever be understood and interpreted through systems of human knowledge which are very much socially contingent. At a theoretical level, and as Donovan argues by drawing on assemblage theory (Donovan 2017), this means that the study of, and knowledge about, what are perceived to be non-socially contingent phenomena should not be outside the scope of social scientific theories of disaster causation (Donovan et al. 2023). Simultaneously, though, this does not alter the fact that the processes themselves – mass movements, seismic shaking and so on – may be entirely non-human and non-social in their specific unfolding. The latter point, and wider questions of landslide causation, is complicated further by the question of anthropogenic climate change which is having some bearing on the rainfall element of the landslide equation (Sharma et al. 2024). In any case, and to return to the original point, one can see that even before we arrive at discussions of social vulnerability, the question of what is or is not natural and/or social about hazards escapes neat distinctions between socially constructed and natural, and thus conceptual, frameworks which rely on such a distinction to function.

While missing the social nature of knowledge about hazards outlined above, an entirely constructivist approach may also leave scholars blind to the fact that the political, economic and cultural processes normally associated with the study of vulnerability are equally applicable to the unfolding of landslide hazards. In their highly cited review of the trends in landslide occurrence in Nepal, Petley et al. (2007) outline in depth how the physical geography of the Himalayas continues to cause landslides there. However, they also point out that the uptick in frequency of landslides in Nepal over the decades preceding their study could not be explained solely by any significant change to the geophysical processes underpinning landslide causation. Instead, they point out that anthropogenic activities like slope cutting for road construction, urbanization, poor drainage systems and dynamite blasting for tunnels are also factors which might explain the increase in landslide activity (see also Froude and Petley 2018). In a more recent study, Rosser et al. (2021) are more explicit in their assertion that a recent uptick in landslide activity following the 2015 Gorkha earthquake cannot be explained solely by the slope instability caused by earthquake shaking. Indeed, their modelling of the expected number of landslides caused by the earthquake is only half the actual number. To explain this, they point to the fact that, in 2015, a new constitution was signed in the country following years of latent conflict, paving the way for a new government. Then, in 2017, there was a general election in which road connectivity became a key bargaining tool for politicians standing for election (Gurung 2021). This led to a road construction boom during and after the election across Nepal, following a thirty-year period in which road construction tripled nationwide – a rise that has been, and continues to be, at least partly geopolitically motivated, mostly related to border tensions between India and China (Murton 2017; Paudel and Le Billon 2020). All of this suggests that landslide hazards in Nepal are often not 'natural hazards' because they are frequently caused in part by socially contingent processes such as road construction, drainage management and border maintenance (Robson et al. 2021). However, it would be equally difficult to argue that landslide hazards are 'socially constructed' given their causation also relies on non-socially contingent processes related to Himalayan geology, the collisions of tectonic plates and the south-west monsoon. Thus, researchers analysing landslide hazards in the Himalayas are as likely to encounter the geopolitical as the geological in their analysis and are turning to analytical frameworks that can account for this.

Some scholars have turned to assemblage-thinking to plug this analytical gap and link analyses of 'hazards' to processes of development that generate conditions of vulnerability and ultimately define what comes to be known as disasters in the Himalayas (Vaishnava and Baka 2021) and beyond (Bobbette and Donovan 2019; Donovan 2020). A paper from my doctoral research on landslide hazards across the border from Nepal in Kalimpong, India, provides an example of this (McGowran 2024a). Rather than separating hazards and vulnerabilities from the outset, the assemblage approach identifies complex interactions of discursive and material processes within 'geopolitical assemblages', which tie together discursive framings of the region as a 'strategic location' with material objects like roads, dams and railway tunnels. This analysis aims to understand how these assemblages work to make disasters possible and then prolong and exacerbate their impacts. In so doing, my analysis moves debates about disasters beyond their specific nature towards a more grounded perspective on what disasters and disaster management mean on the ground in Kalimpong. This helped me highlight disconnects between how disasters are experienced and managed in Kalimpong and what the policy implications of this disconnect are. A tangible recommendation emerging from this analysis is that (landslide) disaster risk reduction in the region should focus on ensuring that infrastructure development, such as roads, railways and drainage, is risk informed. At a more theoretical level, the analysis reflects that achieving any significant policy change in disaster risk reduction would require unsettling entrenched power relationships and rethinking the time frames through which disasters are imagined and thus 'managed' (McGowran 2024b).

In this section, I have argued that assemblage-thinking can take those engaged with critical analyses of disasters beyond the society–nature binary and claims of disasters being 'socio-natural' in origin. Using the example of landslides in the Himalayas, I argued that an approach attempting to distinguish between what is socially constructed and what is natural is unlikely to grasp how and why disasters unfold in specific places. I have argued that assemblage-thinking can aid those of us working in critical disaster studies to reframe the question of nature in disasters so that the statement 'There is no such thing as a natural disaster' comes to be seen not as a conclusion but as a place to start.

Reassembling Disaster Temporalities

As alluded to in the above section, critical studies of disasters are beginning to move beyond claims that disasters are socio-natural processes. In place of these studies, there appears to have been a recent uptick in scholarly work on the temporalities of disasters (Cloke et al. 2017; Cutter 2018; Bosher et al. 2021; Usón and Stehrenberger 2021; Coates 2022; Coates and Warner 2023). In short, scholars have found that in trying to answer the question of 'what a disaster is' (assuming it is not natural), it is useful to define when disasters begin and end. As it turns out, this is no easy task (Remes et al. 2021). This recent uptick notwithstanding, geographers have been critically reflecting on and unravelling the temporalities of disasters for decades. In this section, I will argue that using geographically inflected assemblage-thinking to investigate the temporalities of what come to be known as disasters can help reveal how, paradoxically, the power dynamics that underpin their emergence might also be enrolled in attempts to manage them.

What might be described as a 'temporal turn' in the study of disasters could be linked to several factors. Some might argue that this is a case of 'old wine in new bottles', pointing to the established literature on disasters and development and the main argument emerging from it that disasters reflect, or are simply part of, long-term development processes (Cuny 1983; Collins 2009, 2018). They might also cite Hewitt's (1983) concept of the 'disaster archipelago', a critique of the idea that disasters are 'exceptional' and should be studied as such, as evidence of this reinvention of the wheel. Others may argue that the temporal turn results in geography's ongoing influence on the study of hazards, given a similar rise in critical geographical work on temporalities (Datta 2020; Kitchin 2023) – itself grounded in a much longer history of scholarship on the governance of emergencies and crises (Collier and Lakoff 2008; Anderson 2010; Amoore 2013). Critical human geographers have long deconstructed temporal narratives of immediacy and urgency during periods of crisis, emergency and disaster to show that they often serve to justify and maintain entrenched power relationships (O'Grady 2018). Such a line of critique is often inspired by readings of Agamben and his concept of the 'state of exception' (Agamben 2004), which is in turn underpinned by a Foucauldian understanding of states and governmentality. There are also traces of other disciplinary perspectives in the temporal turn on disasters, spanning critical urbanism (Coates 2018; Cheek and Chmutina 2022), sociology (Guggenheim 2014; Coates and Warner 2023), history (Knowles 2020), anthropology (Oliver-Smith 1999a,b; Zeiderman 2016) and emergency management (Montano and Savitt 2023).

Beyond the influence of specific disciplines, it would also be reasonable to argue that the COVID-19 pandemic played some part in bringing about a 'temporal turn' in disaster studies. Not only did the pandemic disrupt and strain the conventional temporal imaginaries through which emergencies have typically been understood and governed (Grove et al. 2021: 1), but it was also somewhat unique in that it was a disaster experienced in some capacity by most scholars and academics worldwide. It seems likely that this would have had some effect on shaping or co-producing scholars' personal perspectives on what a disaster is and how it should be analysed and written about. In that sense, the pandemic surely set in motion new dynamics in the ongoing co-production of science and society (Jasanoff 2004), some of which may become clearer once the period is analysed in retrospect by historians in the future (Horowitz and Remes 2021).

Disaster management in India provides a useful case study to highlight the productive potential of critically analysing the temporalities of disasters and disaster management using geographical and assemblage-based theoretical approaches. In India, national and state policies on disaster management are largely organized around responding to large-scale, short-term and high-fatality disasters (Seidler et al. 2018). Scholars often point to events that occurred around the time of the establishment of the 2005 National Disaster Management Act as important in this process of territorialization: the Odisha supercyclone in 1999 (death toll: 15,000–30,000 depending on source), the Gujarat earthquake in 2001 (death toll: approx. 13,000–20,000) and the Indian Ocean tsunami of 2004 (death toll in India: 10,000+) (Nambiar 2015; Seidler and Bawa 2016). The introduction and development of this act was supported by international policy frameworks such as the UNDRR Sendai Framework and associated funding mechanisms such as the World Bank's Global Facility for Disaster Risk Reduction. The data- and technology-focused nature of such frameworks and funding mechanisms has led to increased research and funding for the development of (early) warning systems, the streamlining of response mechanisms and efforts towards prevention of death (The World Bank 2021). These improvements are uneven and rely largely on subnational political action, evidenced, for example, by a brief analysis of cyclone disaster risk management in the state of Odisha.

In 1999, Category 5, 'Odisha supercyclone', made landfall on the coast of the state and contributed to approximately 10,000 deaths, though estimates vary. Fourteen years later in 2013, Category 4, Cyclone Phallin, struck the state, leading to a considerable but significantly smaller death toll (less than 100) (Jha et al. 2015; Ray-Bennett 2018) amid widespread economic and infrastructural damage. Jha et al. (2015) attribute this reduction mostly to sustained political and economic investments in preparedness measures and the creation of systems and frameworks which facilitate the rapid mobilization of response and relief measures. There are temporal dynamics here in that the most well-established (and funded) aspects of disaster management tend to relate to short-term actions and mechanisms that are enacted in the hours and days before the hazard materializes (warning systems) and the hours and days following the hazard occurrence (response and relief). The relative success of these measures has informed policy logics on how disasters ought to be managed in India. However, it has also been argued that the focus on early warning and improving response mechanisms through advances in science and technology has rendered India's approach to and framing of disaster management relatively ineffective in addressing disasters that do not fit the temporal profiles of events such as cyclones and earthquakes (McGowran 2024b). For example, staying with Odisha, this means that while responses to large-scale events like cyclones have improved, deaths due to smaller hazards such as lightning strikes have increased. Similarly, efforts to mitigate long-term infrastructural and property damage have not kept pace with the prevention of deaths (Ray-Bennett 2018).

These dynamics arose as a key theme in a study I was part of that qualitatively investigated the impacts of the COVID-19 pandemic and lockdown on migrant workers in India. Drawing from assemblage theory and related work on disaster imaginaries (e.g. Walshe et al. 2023), we (McGowran et al. 2023a) found that the disconnect between the imagined temporality of COVID-19 (as a disaster) and what actually happened was a key factor behind some of the major problems with India's pandemic response: a response that created an internal migration crisis and led to between 3,710,000 (COVID-19 Excess Mortality Collaborators 2022) and 4,740,894 (WHO 2022) excess deaths. We pointed out that, in the early days of the pandemic, the central government commissioned the 'COVID-19 India National Supermodel': a kind of 'warning system' based on epidemiological models which they thought would allow them to mobilize resources in anticipation of localized waves of infection (Department for Science and Technology 2020), despite the explicitly acknowledged lack of reliable data (Agrawal et al. 2021). Indeed, success with similarly structured responses to cyclones was explicitly cited as evidence of why this approach should be taken (Department for Science and Technology 2020: n.p.). Besides the fact that the spatial scale of the pandemic massively exceeded the government's ability to collect data on cases and effectively model (Sinha 2021), this approach was also undermined by the temporal scale of the pandemic. As a government official noted in response to a question about ex gratia payments to the families of people who died in the pandemic (National Disaster Management Authority 2021), 'the disaster had not abated' and ultimately exceeded the temporal imagination of disasters that India's Disaster Management Act and Authority have been structured around. Ultimately, by the time the lethal second wave of the pandemic hit the country, resources, efforts, policies and frankly the political will to deal with the pandemic had all appeared to dwindle (Sharma 2022; Jaffrelot, n.d.). The virus was effectively left to 'let rip' and so it did. Of course, these temporal mismatches can only partly explain the limitations of India's pandemic response. However, they are certainly an important aspect of this explanation. Indeed, while the assemblage approach we took attended to these questions of temporality, it also incorporated questions of nature, space, scale, labour and ultimately power into our analysis of India's response to the pandemic (as a disaster). Perhaps most crucially, the analysis supported arguments made by others (Sengupta and Jha 2021; Singh 2022) that the approach taken and the temporalities inherent to it are inextricably linked to the central government of India's ongoing approach to monopolize political power at the centre within what is purportedly a federal system.

The process of disaster management is at base a process through which temporal imaginations of disasters and the material responses they co-produce come to be projected onto and woven into space and place (Vergara-Pinto and Marín 2023). These processes can make things worse, as happened in India during the pandemic (Raju and Ayeb-Karlsson 2020; Sengupta and Jha 2020), and at the very least produce political contestation. Disaster temporalities can serve as interesting entry points to analyse these dynamics and power relations and ultimately help to explain why governmental policies and practices do not generally respond to people's needs in the places where disasters happen or to the actual time frames in which people experience the impacts of disasters (Walshe et al. 2020; Staupe-Delgado and Rubin 2022). Thus, analysing disaster temporalities often leads back to questions about the relationships between disasters and development and between the everyday and the exceptional. For example, returning to the research I was involved with on the COVID-19 pandemic in India, specifically the southern state of Kerala, our assemblage-based empirical analysis concludes that the COVID-19 pandemic brought into sharp focus the everyday conditions of vulnerability experienced by domestic migrant workers (McGowran et al. 2023b). In turn, the analysis problematized responses to try and address these existing conditions through temporary and exceptional measures under the auspices of disaster management, as opposed to taking more long-term steps to tackle the more permanent causes of their marginalization, the latter being found to be caused by various factors including xenophobic depictions of the 'foreign-ness' of domestic migrants, insecure labour facilitated by labour contractors and administrative barriers to their receipt of welfare payments related to the politics and administration of India's federal system of government. This assemblage-based analysis of disaster temporalities provided a sharp lens for them to analyse and critique the paradoxical and 'schizophrenic' (Lavell and Maskrey 2014: 278) attempts of those in power to manage problems which their own decisions have periodically made worse.

Power and Knowledge in Disaster Assemblages

While initially acknowledging that a consideration of disaster natures can be an important aspect of an assemblage-based analysis of disaster geographies, in the previous section I suggested that disaster temporalities may serve as more productive analytical entry points. A temporal approach can be particularly useful when trying to identify and highlight the entrenched power relations that both make possible and perpetuate conditions of vulnerability: conditions that often become framed as disastrous once they enter relations with hazards and attempts to mitigate them. It is often the case that such analyses of these power relations end up leading to critical reflections on the co-production of power and knowledge as it relates to disasters (Pelling et al. 2021; Gurung and McGowran 2022; Vergara-Pinto and Marín 2023). Indeed, it has been argued that assemblage-based approaches are uniquely well positioned to facilitate such reflections (Donovan et al. 2023).

Rather than getting bogged down in debating whether disasters are events or processes, Usón and Stehrenberger (2021) draw on an assemblage-based approach to argue that disasters might be better understood as 'temporal devices' in their own right, used to frame processes as events and events as processes according to the political persuasion of the framer(s). Using the example of the 1970 Ancash earthquake in Peru, they suggest that while it is correct to point out that risk of the disaster was 'constructed' over several years (Oliver-Smith 1999a), this framing has also been used to imply blame and responsibility on specific actors and institutions. In essence, they argue that regardless of whether it is viewed as an event or a process, the narrative framings of the temporalities of the earthquake have been used to bolster a variety of political narratives: spanning the optimistic and the pessimistic, and the transformational and the conservative. This points to the fact that framing hazard-related disasters as socially constructed is immediately political: it implies blame, responsibility, liability and so on. This is very often necessary. More often than not, though, blame and responsibility are difficult to attribute to one or a few choices, let alone individuals. Thus, while it is often argued that framing disasters as social constructs is a way to hold to account those who preside over the decisions which make disasters worse, the opposite can also be true. Through carefully constructed rhetoric and narrative framing, it is entirely possible for those with power to agree that a disaster was of course related to power and politics and largely socially constructed (Angell 2014). Far from challenging them, many examples show that this usually leads those in power to use the disaster to justify remaining in power, and even extending it, so that disasters can be managed in an improved way in the future (Pelling and Dill 2010; Kelman 2011). Indeed, and in relation to the discussion thus far, constructing narrative framings of disasters with respect to their nature and temporality is central to this process of power maintenance.

To that end, Usón and Stehrenberger (2021) also make the point that scholarly studies are often themselves enrolled in the temporal framing of disasters and thus caution disaster researchers from all disciplinary persuasions to critically reflect on that fact. In turn, they point to the utility of assemblage-thinking in achieving those ends and are certainly not the only ones to have done so (Pelling et al. 2021; Vaiciulyte et al. 2023; Vergara-Pinto and Marín 2023). For that reason, I suggest that alongside analyses of disaster natures and temporalities, assemblage-based analyses of disaster geographies are well placed to provide critical reflections on the intersections of power and knowledge which hold disaster geographies in place. This process of critical self-reflection would be an important process alongside many others (Gaillard 2019; Meriläinen et al. 2021; Yadav et al. 2022) in any attempt to reconcile the disconnects between academic, scientific and/or some 'Western' knowledge systems with other knowledge systems that might provide more grounded and useful information about 'what we call disasters' (Gaillard 2021) in the places they actually occur. As stressed thus far, thinking about how these disconnects and crossovers in knowledge systems cohere around the themes of nature, temporality and power might be a productive framework for negotiating, working with and making useful the knowledge that emerges from disaster research. A similar point could be made about the need for critical reflection on knowledge and power in interdisciplinary disaster research (Donovan et al. 2023), where power inequalities may not be as complex and pronounced as, for example, in research by European researchers working in postcolonial contexts.

Conclusion

This chapter was arranged around three interconnected themes that might give shape to assemblage-based analyses of disaster geographies. The first and second revolved around disaster natures and disaster temporalities, respectively. The third combined questions of nature and time in a wider reflection on disaster politics. In closing, I call for research that combines and thinks through the connections between these three themes. For example, thinking through the temporalities of disasters might provoke critical reflections on the scalar and spatial arrangements of disaster risk management apparatuses, policy frameworks and research practices. In turn, analyses of the latter might also critically reflect on the nature of the problems such apparatuses and policy frameworks are and become concerned with, and how these do or do not correspond with people's experiences of these problems. Assemblage-thinking can help to unearth and problematize key points of connection and disconnection between disasters and attempts to manage them. In so doing, and specifically by detailing how these points of (dis)connection vary across spatial and temporal scales, such analyses can provide nuanced and grounded critiques of disaster (risk) management while provoking new ways of thinking about and addressing problems that come to be known and managed as disasters.

To that end, assemblage-based approaches also provoke scholars to critically reflect on the ontologies underpinning their approaches to studying disasters. Such reflections represent an important first step in facilitating more interdisciplinary ways of working across the physical and social sciences, and geographical contexts, in disaster studies and geography. They may also provoke critical reflections on how, where and why knowledge of disasters is constructed. Thus, assemblage-based research might also support, nuance and challenge where necessary increasingly influential scholarly turns towards disaster research that seeks to challenge established systems of knowledge production.

References

Agamben, G. (2004), *State of Exception*, University of Chicago Press.

Agrawal, M., M. Kanitkar and M. Vidyasagar (2021), 'Modelling the Spread of SARS -CoV-2 Pandemic – Impact of Lockdowns & Interventions', *Indian Journal of Medical Research* 153 (1–2): 175–81.

Alcántara-Ayala, I., I. Burton, A. Lavell, E. Mansilla, A. Maskrey, A. Oliver-Smith and F. Ramírez-Gómez (2021), 'Root Causes and Policy Dilemmas of the COVID-19 Pandemic Global Disaster', *International Journal of Disaster Risk Reduction* 52:101892.

Amoore, L. (2013), *The Politics of Possibility: Risk and Security Beyond Probability*, Duke University Press.

Anderson, B. (2010), 'Security and the Future: Anticipating the Event of Terror', *Geoforum* 41 (2): 227–35.

Anderson, B. and C. McFarlane (2011), 'Assemblage and Geography', *Area* 43 (2): 124–7. https://doi.org/10.1111/j.1475-4762.2011.01004.x.

Angell, E. (2014), 'Assembling Disaster: Earthquakes and Urban Politics in Istanbul', *City* 18 (6): 667–78. https://doi.org/10.1080/13604813.2014.962881.

Blaikie, P. M. and H. C. Brookfield (1987), *Land Degradation and Society*, Methuen.

Bobbette, A. and A. Donovan (2019), 'Political Geology: An Introduction', in A. Bobbette and A. Donovan (eds), *Political Geology: Active Stratigraphies and the Making of Life*, 1–34, Springer.

Bosher, L., K. Chmutina and D. van Niekerk (2021), 'Stop Going Around in Circles: Towards a Reconceptualisation of Disaster Risk Management Phases', *Disaster Prevention and Management: An International Journal* 30 (4/5): 525–37. https://doi.org/10.1108/DPM-03-2021-0071.

Braun, B. (2005), 'Environmental Issues: Writing a More-than-Human Urban Geography', *Progress in Human Geography* 29 (5): 635–50.

Buchanan, I. (2000), *Deleuzism: A Metacommentary*, Edinburgh University Press.

Castree, N. (2005), *Nature*, Routledge.

Castree, N. and B. Braun (2001), *Social Nature: Theory, Practice, and Politics*, Blackwell.

Cheek, W. and K. Chmutina (2022), 'Measuring Resilience in the Assumed City', *International Journal of Disaster Risk Science* 13 (3): 1–13.

Cloke, P., S. Dickinson and S. Tupper (2017), 'The Christchurch Earthquakes 2010, 2011: Geographies of an Event', *New Zealand Geographer* 73 (2): 69–80.

Coates, R. (2018), 'Exploring Disaster Eventfulness in Urbanizing Brazil', *Peace Review* 30 (2): 144–51. https://doi.org/10.1080/10402659.2018.1458941.

Coates, R. (2022), 'Infrastructural Events? Flood Disaster, Narratives and Framing Under Hazardous Urbanisation', *International Journal of Disaster Risk Reduction* 74:102918. https://doi.org/10.1016/j.ijdrr.2022.102918.

Coates, R. and J. Warner (2023), 'Calamitous Events? Exploring Perceptions of Disaster', *International Journal of Disaster Risk Reduction* 91:103700. https://doi.org/10.1016/j.ijdrr.2023.103700.

Collier, S. J. and A. Lakoff (2008), 'Distributed Preparedness: The Spatial Logic of Domestic Security in the United States', *Environment and Planning D: Society and Space* 26 (1): 7–28.

Collins, A. E. (2009), *Disaster and Development*, Routledge.

Collins, A. E. (2018), 'Advancing the Disaster and Development Paradigm', *International Journal of Disaster Risk Science* 9 (4): 486–95. https://doi.org/10.1007/s13753-018-0206-5.

COVID-19 Excess Mortality Collaborators (2022), 'Estimating Excess Mortality Due to the COVID-19 Pandemic: A Systematic Analysis of COVID-19-Related Mortality, 2020–21', *Lancet* 399 (10334): 1513–36. https://doi.org/10.1016/S0140-6736(21)02796-3.

Cuny, F. C. (1983), *Disaster and Development*, Oxford University Press.

Cutter, S. L. (2018), 'Compound, Cascading, or Complex Disasters: What's in a Name?', *Environment: Science and Policy for Sustainable Development* 60 (6): 16–25.

Datta, A. (2020), 'The "Smart Safe City": Gendered Time, Speed, and Violence in the Margins of India's Urban Age', *Annals of the American Association of Geographers* 110 (5): 1318–34. https://doi.org/10.1080/24694452.2019.1687279.

DeLanda, M. (2016), *Assemblage Theory*, Edinburgh University Press.

Deleuze, G. and C. Parnet (1987), *Dialogues*, trans. Hugh Tomlinson and Barbara Habberjam, Columbia University Press.

Demeritt, D. (2002), 'What Is the "Social Construction of Nature"? A Typology and Sympathetic Critique', *Progress in Human Geography* 26 (6): 767–90.

Department of Science and Technology (DST) (2020), 'DST Initiates COVID-19 India National Supermodel for Monitoring Infection Transmission & Aid Decision-Making by Policymakers'. Available at: https://dst.gov.in/dst-initiates-covid-19-india-national-supermodel-monitoring-infection-transmission-aid-decision (accessed 7 November 2025).

Donovan, A. (2017), 'Geopower: Reflections on the Critical Geography of Disasters', *Progress in Human Geography* 41 (1): 44–67. https://doi.org/10.1177/0309132515627020.

Donovan, A. (2020), 'When the Vertical Becomes Horizontal: Experiencing Exploding Mountains in Borderlands', *Annals of the American Association of Geographers* 110 (4): 1040–58. https://doi.org/10.1080/24694452.2019.1677213.

Donovan, A., J. Morin and R. Walshe (2023), 'Progress Report: Interdisciplinary Research in Hazards and Disaster Risk', *Progress in Environmental Geography* 2 (3): 202–22. https://doi.org/10.1177/27539687231183448.

Fraser, A., S. Patterson and M. Pelling (2014), 'A Review of the FORIN Methodology and Existing FORIN Case Studies', *Journal of Extreme Events* 3 (2): 1650008. https://doi.org/10.1142/S2345737616500081.

Froude, M. J. and D. N. Petley (2018), 'Global Fatal Landslide Occurrence from 2004 to 2016', *Natural Hazards and Earth System Sciences* 18 (8): 2161–81.

Gaillard, J. C. (2019), 'Disaster Studies Inside Out', *Disasters* 43 (S1): S7–S17. https://doi.org/10.1111/disa.12323.

Gaillard, J. C. (2021), *The Invention of Disaster: Power and Knowledge in Discourses on Hazard and Vulnerability*, Routledge.

Grove, K. (2009), 'Rethinking the Nature of Urban Environmental Politics: Security, Subjectivity, and the Non-Human', *Geoforum* 40 (2): 207–16.

Grove, K., L. Rickards, B. Anderson and M. Kearnes (2021), 'The Uneven Distribution of Futurity: Slow Emergencies and the Event of COVID-19', *Geographical Research* 60 (1): 6–17. https://doi.org/10.1111/1745-5871.12501.

Guggenheim, M. (2014), 'Introduction: Disasters as Politics – Politics as Disasters', *Sociological Review* 62 (1_suppl): 1–16.

Gurung, L. and P. McGowran (2022), 'Local Perspectives on Landslide Prevention and Management in Kalimpong District, West Bengal, India', *Disaster Prevention and Management: An International Journal* 31 (3): 193–201. https://doi.org/10.1108/DPM-02-2021-0056.

Gurung, P. (2021), 'Challenging Infrastructural Orthodoxies: Political and Economic Geographies of a Himalayan Road', *Geoforum* 120:103–12. https://doi.org/10.1016/j.geoforum.2021.01.020.

Hewitt, K. (1983), *Interpretation of Calamity: From the Viewpoint of Human Ecology*, Allen & Unwin.

Heynen, N., M. Kaika and E. Swyngedouw (2006), *In the Nature of Cities: Urban Political Ecology and the Politics of Urban Metabolism*, Taylor & Francis.

Horowitz, A. and J. A. C. Remes (2021), 'Introducing Critical Disaster Studies', in A. Horowitz and J. A. C. Remes (eds), *Critical Disaster Studies*, 1–8, University of Pennsylvania Press. https://doi.org/10.9783/9780812299724-001.

Jaffrelot, C. (n.d.), 'India's Second Wave: A Man-Made Disaster?', *Institut Montaigne*. Available at: https://www.institutmontaigne.org/en/expressions/indias-second-wave-man-made-disaster (accessed 9 August 2024).

Jasanoff, S. (2004), *States of Knowledge: The Co-production of Science and the Social Order*, Routledge.

Jha, A., R. Basu and A. Basu (2015), 'Studying Policy Changes in Disaster Management in India: A Tale of Two Cyclones', *Disaster Medicine and Public Health Preparedness* 10 (1): 42–6. https://doi.org/10.1017/dmp.2015.116.

Kelman, I. (2011), *Disaster Diplomacy: How Disasters Affect Peace and Conflict*, Routledge.

Kelman, I. (2018), 'Islandness Within Climate Change Narratives of Small Island Developing States (SIDS)', *Island Studies Journal* 13 (1): 149–66.

Kelman, I., J. C. Gaillard, J. Lewis and J. Mercer (2016), 'Learning from the History of Disaster Vulnerability and Resilience Research and Practice for Climate Change', *Natural Hazards* 82 (1): 129–43.

Kitchin, R. (2023), *Digital Timescapes: Technology, Temporality and Society*, John Wiley & Sons.

Knowles, S. G. (2020), 'Slow Disaster in the Anthropocene: A Historian Witnesses Climate Change on the Korean Peninsula', *Daedalus* 149 (4): 192–206. https://doi.org/10.1162/daed_a_01827.

Latour, B. (2005), *Reassembling the Social: An Introduction to Actor-Network Theory*, Oxford University Press.

Lave, R. (2015), 'Reassembling the Structural: Political Ecology and Actor-Network Theory', in T. Perreault, G. Bridge and J. McCarthy (eds), *The Routledge Handbook of Political Ecology*, 213–23, Routledge.

Lavell, A. and A. Maskrey (2014), 'The Future of Disaster Risk Management', *Environmental Hazards* 13 (4): 267–80. https://doi.org/10.1080/17477891.2014.935282.

Law, J. (1992), 'Notes on the Theory of the Actor-Network: Ordering, Strategy, and Heterogeneity', *Systems Practice* 5 (4): 379–93.

Legg, S. (2011), 'Assemblage/Apparatus: Using Deleuze and Foucault', *Area* 43 (2): 128–33.

Marston, S. A., J. P. Jones III and K. Woodward (2005), 'Human Geography Without Scale', *Transactions of the Institute of British Geographers* 30 (4): 416–32.

McFarlane, C. (2011), 'Assemblage and Critical Urbanism', *City* 15 (2): 204–24.

McFarlane, C. and B. Anderson (2011), 'Thinking with Assemblage', *Area* 43 (2): 162–4.

McGowran, P. (2024a), 'Geopolitical Assemblages and Disasters-in-the-Making in Kalimpong, India', *Political Geography* 108:103008. https://doi.org/10.1016/j.polgeo.2023.103008.

McGowran, P. (2024b), 'Landslide Disasters in Kalimpong, India: Matters of Time?', *Geoforum* 156:104140. https://doi.org/10.1016/j.geoforum.2024.104140.

McGowran, P., H. Johns, E. Raju and S. Ayeb-Karlsson (2023a), 'The Making of India's COVID-19 Disaster: A Disaster Risk Management (DRM) Assemblage Analysis', *International Journal of Disaster Risk Reduction* 93:103797. https://doi.org/10.1016/j.ijdrr.2023.103797.

McGowran, P., M. A. Mathews, H. Johns, M. C. Harasym, E. Raju and S. Ayeb-Karlsson (2023b), 'Investigating the Conditions of Vulnerability Experienced by Migrant Workers During the COVID-19 Pandemic in Kerala, India', *Disasters* 48:e12614. https://doi.org/10.1111/disa.12614.

Meriläinen, E., J. Joseph, M. Jauhola, P. Yadav, E. Romo-Murphy, J. Marin and S. Gadhavi (2021), 'Examining Relational Social Ontologies of Disaster Resilience: Lived Experiences from India, Indonesia, Nepal, Chile and Andean Territories', *Disaster Prevention and Management: An International Journal* 31 (3): 273–87. https://doi.org/10.1108/DPM-02-2021-0057.

Mol, A. and J. Law (1994), 'Regions, Networks and Fluids: Anaemia and Social Topology', *Social Studies of Science* 24 (4): 641–71. https://doi.org/10.1177/030631279402400402.

Montano, S. and A. Savitt (2023), 'Revisiting Emergencies, Disasters, & Catastrophes: Adding Duration to the Hazard Event Classification', *International Journal of Mass Emergencies & Disasters* 41 (2–3): 259–78. https://doi.org/10.1177/02807270231211831.

Müller, M. and C. Schurr (2016), 'Assemblage Thinking and Actor-Network Theory: Conjunctions, Disjunctions, Cross-Fertilisations', *Transactions of the Institute of British Geographers* 41 (3): 217–29. https://doi.org/10.1111/tran.12117.

Murdoch, J. (1997), 'Inhuman/Nonhuman/Human: Actor-Network Theory and the Prospects for a Nondualistic and Symmetrical Perspective on Nature and Society', *Environment and Planning D: Society and Space* 15 (6): 731–56.

Murton, G. (2017), 'Making Mountain Places into State Spaces: Infrastructure, Consumption, and Territorial Practice in a Himalayan Borderland', *Annals of the American Association of Geographers* 107 (2): 536–45. https://doi.org/10.1080/24694452.2016.1232616.

Nambiar, M. (2015), 'A Decade of Disaster Risk Management in India', *Economic and Political Weekly* 50 (5): 36–40.

National Disaster Management Authority (2021), 'Guidelines for Ex-Gratia Assistance to Next of Kin of the Deceased by COVID-19' (Issued in Compliance to the Hon'ble Supreme Court Order dated 30.06.2021 in WP (Civil) No. 539 and WP (Civil) 554 of 2021), Government of India, National Disaster Management Authority. Available at: https://ndmindia.mha.gov.in/NDMINDIA-CMS/viewWhatsNewPdf-D627 (accessed 7 November 2025).

O'Grady, N. (2018), *Governing Future Emergencies: Lived Relations to Risk in the UK Fire and Rescue Service*, Springer.

O'Keefe, P., K. Westgate and B. Wisner (1976), 'Taking the Naturalness Out of Natural Disasters', *Nature* 260:566–7.

Oliver-Smith, A. (1999a), 'Peru's Five-Hundred-Year Earthquake: Vulnerability in Historical Context', in A. Oliver-Smith and S. Hoffman (eds), *The Angry Earth: Disaster in Anthropological Perspective*, 88–102, Routledge.

Oliver-Smith, A. (1999b), '"What Is a Disaster?": Anthropological Perspectives on a Persistent Question', in A. Oliver-Smith and S. Hoffman (eds), *The Angry Earth: Disaster in Anthropological Perspective*, 32–48, Routledge.

Oliver-Smith, A., I. Alcántara-Ayala, I. Burton and A. Lavell (2017), 'The Social Construction of Disaster Risk: Seeking Root Causes', *International Journal of Disaster Risk Reduction* 22:469–74. https://doi.org/10.1016/j.ijdrr.2016.10.006.

Paudel, D. and P. Le Billon (2020), 'Geo-Logics of Power: Disaster Capitalism, Himalayan Materialities, and the Geopolitical Economy of Reconstruction in Post-Earthquake Nepal', *Geopolitics* 25 (4): 838–66. https://doi.org/10.1080/14650045.2018.1533818.

Pelling, M., H. Adams, G. Adamson, A. Barcena, S. Blackburn, M. Borie, A. Donovan, A. Ogra, F. Taylor and L. Yi (2021), 'Building Back Better from COVID-19: Knowledge, Emergence and Social Contracts', *Progress in Human Geography* 46 (1): 121–38. https://doi.org/10.1177/03091325211059569.

Pelling, M. and K. Dill (2010), 'Disaster Politics: Tipping Points for Change in the Adaptation of Sociopolitical Regimes', *Progress in Human Geography* 34 (1): 21–37.

Petley, D. N., G. J. Hearn, A. Hart, N. J. Rosser, S. A. Dunning, K. Oven and W. A. Mitchell (2007), 'Trends in Landslide Occurrence in Nepal', *Natural Hazards* 43 (1): 23–44.

Raju, E. and S. Ayeb-Karlsson (2020), 'COVID-19: How Do You Self-Isolate in a Refugee Camp?', *International Journal of Public Health* 65 (5): 515–17. https://doi.org/10.1007/s00038-020-01381-8.

Ray-Bennett, N. S. (2018), 'Disasters, Deaths, and the Sendai Goal One: Lessons from Odisha, India', *World Development* 103:27–39. https://doi.org/10.1016/j.worlddev.2017.10.003.

Remes, J. A. C., A. D. Horowitz and A. Horowitz (2021), *Critical Disaster Studies*, University of Pennsylvania Press.

Robbins, P. and B. Marks (2010), 'Assemblage Geographies', in S. J. Smith, R. Pain, S. A. Marston and J. Jones (eds), *The SAGE Handbook of Social Geographies*, 176–94, SAGE.

Robson, E., D. Milledge, E. Robson and S. Utili (2021), 'Road Slope Stabilisation in Nepal: Stakeholder Perspectives', in *XIII International Symposium on Landslides*, International Society for Soil Mechanics and Geotechnical Engineering, Cartagena, Colombia. Available at: https://www.issmge.org/publications/publication/road-slope-stabilisation-in-nepal-stakeholder-perspectives (accessed 7 November 2025).

Rosser, N., M. Kincey, K. Oven, A. Densmore, T. Robinson, D. S. Pujara, R. Shrestha, J. Smutny, K. Gurung, S. Lama and M. R. Dhital (2021), 'Changing Significance of Landslide Hazard and Risk After the 2015 Mw 7.8 Gorkha, Nepal Earthquake', *Progress in Disaster Science* 10:100159. https://doi.org/10.1016/j.pdisas.2021.100159.

Seidler, R. and K. S. Bawa (2016), 'Ancient Risks, Current Challenges in the Himalayas', *Economic and Political Weekly* 51 (41): 63–7.

Seidler, R., K. Dietrich, S. Schweizer, K. S. Bawa, S. Chopde, F. Zaman, A. Sharma, S. Bhattacharya, L. P. Devkota and S. Khaling (2018), 'Progress on Integrating Climate Change Adaptation and Disaster Risk Reduction for Sustainable Development Pathways in South Asia: Evidence from Six Research Projects', *International Journal of Disaster Risk Reduction* 31:92–101. https://doi.org/10.1016/j.ijdrr.2018.04.023.

Sengupta, S. and M. K. Jha (2020), 'Social Policy, COVID-19 and Impoverished Migrants: Challenges and Prospects in Locked Down India', *International Journal of Community and Social Development* 2 (2): 152–72. https://doi.org/10.1177/2516602620933715.

Sengupta, S. and M. K. Jha (2021), 'Risks and Resilience: COVID-19 Response and Disaster Management Policies in India', *India Review* 20 (2): 121–41. https://doi.org/10.1080/14736489.2021.1895568.

Sharma, A., H. Sajjad, N. Bhuyan, Md H. Rahaman and R. Ali (2024), 'Climate Change-Induced Landslide Vulnerability: Empirical Evidence from Shimla District, Himachal Pradesh, India', *International Journal of Disaster Risk Reduction* 110:104657. https://doi.org/10.1016/j.ijdrr.2024.104657.

Sharma, S. D. (2022), 'India's Fight Against the COVID-19 Pandemic: Lessons and the Way Forward', *India Quarterly* 78 (1): 9–27. https://doi.org/10.1177/09749284211068470.

Singh, U. K. (2022), 'Federalism, Democracy and the National Security State in India', *Territory, Politics, Governance* 10 (1): 51–66. https://doi.org/10.1080/21622671.2021.1899975.

Sinha, S. (2021), 'Modelling the Spread of SARS-CoV-2 Pandemic – Is This Even Close to a Supermodel?', *Indian Journal of Medical Research* 153 (1&2): 201–6. https://doi.org/10.4103/ijmr.IJMR_4525_20.

Smith, N. (2010), *Uneven Development: Nature, Capital, and the Production of Space*, University of Georgia Press.

Smith, N. and P. O'Keefe (1980), 'Geography, Marx and the Concept of Nature', *Antipode* 12 (2): 30–9.

Soja, R. and L. Starkel (2007), 'Extreme Rainfalls in Eastern Himalaya and Southern Slope of Meghalaya Plateau and Their Geomorphologic Impacts', *Geomorphology* 84 (3): 170–80. https://doi.org/10.1016/j.geomorph.2006.01.040.

Starkel, L. (1979), 'The Role of Extreme Meteorological Events in the Shaping of Mountain Relief', *Geographica Polonica* 41:13–20.

Starkel, L. and S. Sarkar (2014), 'The Sikkim-Darjeeling Himalaya: Landforms, Evolutionary History and Present-Day Processes', in V. Kale (ed.), *Landscapes and Landforms of India* (World Geomorphological Landscapes), 157–64, Springer.

Staupe-Delgado, R. and O. Rubin (2022), 'Challenges Associated with Creeping Disasters in Disaster Risk Science and Practice: Considering Disaster Onset Dynamics', *International Journal of Disaster Risk Science* 13 (1): 1–11. https://doi.org/10.1007/s13753-022-00391-9.

Swyngedouw, E. and N. C. Heynen (2003), 'Urban Political Ecology, Justice and the Politics of Scale', *Antipode* 35 (5): 898–918. https://doi.org/10.1111/j.1467-8330.2003.00364.x.

The World Bank (2021), 'Disaster Risk Management', The World Bank: Country Partnership Framework for India. Available at: https://www.worldbank.org/en/cpf/india/what-we-work/resource-efficient-growth/disaster-risk-management (accessed 29 December 2021).

UNDRR (2015), *The Sendai Framework for Disaster Risk Reduction 2015–2030*, United Nations Office for Disaster Risk Reduction.

Usón, T. J. and C. S. Stehrenberger (2021), 'A Temporal Device: Disasters and the Articulation of (De)acceleration In and Beyond 1970 Ancash's Earthquake', *Res Publica. Revista de Historia de las Ideas Políticas* 24 (3): 467–80. https://doi.org/10.5209/rpub.79245.

Vaiciulyte, S., H. Underhill and E. Reddy (2023), 'Assembling Fire: Beyond Engineering Solutions', *Disaster Prevention and Management: An International Journal* 33 (3): 237–49. https://doi.org/10.1108/DPM-05-2023-0126.

Vaishnava, S. and J. Baka (2021), 'Unruly Mountains: Hydropower Assemblages and Geological Surprises in the Indian Himalayas', *Environment and Planning E: Nature and Space* 5 (3): 1123–45. https://doi.org/10.1177/25148486211050780.

Vergara-Pinto, F. and A. Marín (2023), 'Stratigraphy of Volcanic Memory: Sociocultural Dimensions of Volcanic Risk in the Southern Andes, Chile', *Journal of Contingencies and Crisis Management* 31 (4): 1018–33. https://doi.org/10.1111/1468-5973.12474.

Walshe, R. A., G. C. D. Adamson and I. Kelman (2020), 'Helices of Disaster Memory: How Forgetting and Remembering Influence Tropical Cyclone Response in Mauritius', *International Journal of Disaster Risk Reduction* 50:101901.

Walshe, R., J. Morin, A. Donovan, F. Vergara-Pinto and C. Smith (2023), 'Contrasting Memories and Imaginaries of Lonquimay Volcano, Chile', *International Journal of Disaster Risk Reduction* 97:104003. https://doi.org/10.1016/j.ijdrr.2023.104003.

Watts, M. (1983), *Silent Violence: Food, Famine and Peasantry in Northern Nigeria*, University of California Press.

Whatmore, S. (2002), *Hybrid Geographies: Natures, Cultures, Spaces*, SAGE.

WHO (2022), '14.9 Million Excess Deaths Associated with the COVID-19 Pandemic in 2020 and 2021', World Health Organisation. Available at: https://www.who.int/news/item/05-05-2022-14.9-million-excess-deaths-were-associated-with-the-covid-19-pandemic-in-2020-and-2021 (accessed 12 May 2022).

Wisner, B., P. Blaikie, T. Cannon and I. Davis (2003), *At Risk II: Natural Hazards, People's Vulnerability and Disasters*, Routledge. Available at: https://ebookcentral.proquest.com/lib/kcl/detail.action?docID=182221.

Yadav, P., J. Lassa, V. Marchezini and D. van Niekerk (2022), 'Guest Editorial: Introduction to Calling for Change in Disaster Studies – Rethinking Disaster Studies', *Disaster Prevention and Management: An International Journal* 31 (3): 177–81. https://doi.org/10.1108/DPM-06-2022-418.

Zeiderman, A. (2016), 'Prognosis Past: The Temporal Politics of Disaster in Colombia', *Journal of the Royal Anthropological Institute* 22 (S1): 163–80. https://doi.org/10.1111/1467-9655.12399.

5

How the Production of Economic and Scientific 'Facts' Constrains State Sovereignty

A Postcolonial Critique of Sovereign Debt Disaster Clauses

John Hogan Morris and Sam Simkin

Hurricane Beryl, a Category 5[1] tropical storm, struck the Caribbean on 1 July 2024, demolishing more than 90 per cent of buildings in parts of Grenada and St Vincent and the Grenadines before reaching Jamaica as a Category 4 storm (Duncan 2024). Once there, heavy winds and rain destroyed more than US$6.4 million in food crops and supporting infrastructure. While natural disasters such as Beryl often garner immediate media attention for headline estimates of financial losses, these events have longer-term economic impacts that extend well beyond the time horizons of the television news cycle. Substantial economic losses restrict national governments' abilities to issue government bonds and access credit, thus stymieing recovery and exacerbating their vulnerability to future climate disasters. In an attempt to address these pernicious long-term trends, the Grenadian government responded to Beryl's landfall in Grenada by triggering a 'hurricane clause' to pause US$12 million in interest payments to private investors in its US$112 million sovereign bond (Cotterill and Harris 2024).

This chapter addresses this recent development in disaster planning and response, namely, the inclusion of disaster clauses in the government bonds issued by three Caribbean states: Grenada, Barbados and the Bahamas. These clauses allow countries the option to temporarily suspend debt repayments should they experience weather or climate disasters meeting predefined and contractually agreed parameters such as storm severity or modelled losses caused by the disaster (Jarzabkowski et al. 2023). Such clauses require quantifiable and verifiable indicators of financial losses and/or disaster magnitude. These enable debtors to make the sovereign decision to pause debt repayments to their creditors. Once recovered from the initial economic shock of the disaster, debt relations are normalized and repayments resume.

An exploration of sovereign debt disaster clauses allows us to deepen geographical understandings of disasters. Geographers have employed relational thought to account for the way disasters are frequently managed through networks of diverse entities that often work together despite being individually located in separate places. Despite the strengths of this literature, it appears ill equipped for the analysis of sovereign debt disaster clauses. Disaster scholarship anchored in assemblage theory does not allow an ethical critique of the individual, non-human, calculative components of risk management through disaster clauses. Alternatively, work that analyses financial responses to disasters as post-plantation ecologies identifies the racially hierarchical system of sovereign debt as the context that conditions the sovereignty of postcolonial states. Although the legal and monetary contexts of sovereign debt disaster clauses constrain the sovereignty of disaster-vulnerable states, scientific and computational practices also actively constrain the sovereign decision-making space. In this way, this chapter contributes to an emerging literature on disaster measurement as a political technology (Perry 2024).

In what follows, we analyse the conditions needed to activate these clauses as 'hinge points' for critical inquiry because they are constructed as scientific and economic facts through processes that prevent many circumstances from being eligible for debt repayment suspensions. The number of factors required to trigger a disaster clause influences the likelihood of national governments being presented with the decision to suspend debt repayments. Further, creditors agree to disaster clauses when they have confidence in the objectivity and impartiality of the data used to model losses and to measure the disasters. Across these three bonds, the relevant datasets have been collected using US federal government–funded scientific infrastructures and techniques. The conditions that enable the sovereign decisions of Grenada, Barbados and the Bahamas are subsequently subject to funding by the US government and vulnerable to political decision-making beyond their control. Furthermore, the datasets themselves impose constraints on sovereign decisions, by virtue of the geophysical phenomena that can be measured and the limitations of data. Thus, sovereign decisions are foreclosed for some commonly occurring disasters that are directly linked to historic colonial land use and for disasters in which the data is hard to model due to difficulty in finding proxies for conditions across different scales. In this way, calculation constitutes an enabling condition of postcolonial statecraft.

The Relationality of Disaster Risk Management

As the previous chapter in this volume illustrates, scholarship on disaster-vulnerable spaces and places has already been influenced by the wider 'mainstreaming of relationality' as a prompt for geographical research and debate (Anderson et al. 2012: 185). This yields a conceptualization of space as more than an empty container in which disasters occur. Instead, diverse components are thought to manage disaster risk by working in concert across multiple sites through relational space (Grove 2013, 2014a,b; Grove and Pugh 2015; Donovan 2017; McGowran and Donovan 2021; McGowran et al. 2023; Perry 2023; McGowran 2024a,b, this volume).

The 'analytics of assemblage' draws attention to the role of collective agency in social life and has allowed geographers to capture the way disasters are managed and mitigated through hybrid collectives of 'bodies, institutions, techniques, strategies, and rationalities' (Grove 2013: 570). A significant motivation for this analytic approach is to avoid conceptualizing disasters as either completely natural events or entirely social constructions (see Donovan 2017). 'Disaster risk management assemblages' – such as the disaster itself, the equipment used for data collection, models used for supplementing empirical observation, methods for identifying and valuing elements at risk of a disaster, and the production of risk maps – are therefore analysed as thoroughly material collectives which are simultaneously more-than-human yet not non-human (McGowran and Donovan 2021). Technical and scientific devices are thus viewed as active agents through which future and emergent disasters are understood and governed as 'risks'. McGowran and Donovan (2021: 1610) emphasize the role that 'interactions between' the more-than-human elements of the assemblage play in the actual materialization of disaster risk. Agency and causality are thus 'co-produced by a range of bodies/objects/knowledge as they are brought into relation with one another' (Waterton and Dittmer 2014: 125).

Disaster clauses in sovereign debt may well reveal a limitation of assemblage approaches. Computation and measurement play a decisive role in determining how clause triggers are set and when they are activated. As such, the conditions for the sovereign decision contain critical non-human components and should be amenable to ethical analysis. As Anderson et al. (2012: 186) caution, the analysis of causality as 'an emergent property of intra-action' offered by assemblage-thinking has consequences for the kind of ethical critique that can be offered. McGowran and Donovan (2021: 1611) maintain that analysis of distributed or collective agency can still attend to the 'political, economic and social power relations' from which disasters emerge, but it remains unclear that this provides the foundations for an ethical critique of the non-human components of hybrid collectives.

Other variants of relational thought provide a more direct ethical critique of disaster risk management. Scholarship on disasters in the Caribbean situates the conceptual vocabulary of relational space within broader concerns of racial capitalism and environmental justice (Sheller 2018; Perry 2022, 2023, 2024). The Caribbean's acute exposure to climate change and accrual of recovery-related debt is rooted in 'colonial patterns of land settlement', the exploitation of 'Black and enforced labour' and resultant 'racial hierarchies' that facilitated extraction and accumulation (Perry 2023). Those plantation[2] systems have shaped the contours of today's debt burdens and limited economic diversification. The result is postcolonial states that are dependent on 'environmentally sensitive land, topography and external sources of finance'.

Keston Perry's (2023) analysis of the way debt is used as a response to fiscal and climate disasters in the Caribbean employs ecological thought to illustrate how particular configurations of relational space facilitate contemporary exploitation and extraction. This approach decomposes systems to the level of the different 'ecologies' that constitute them and has previously been employed in financial geography to isolate distinct arrangements of institutions, markets and economic actors and particular configurations of space, often to explain variegated outcomes such as socio-spatial inequalities (Leyshon 2020). For Perry (2023), 'post-plantation ecologies' comprise coercive and hierarchical relations at varying scales which echo colonial power dynamics. Viewed this way, financial instruments that ostensibly mitigate the impact of disasters – such as catastrophe insurance, debt swaps and public debt – subsume disaster-vulnerable societies into racialized financial geographies. The intersection of financial and disaster vulnerability allows previously colonized states to be rated, priced and traded for investors to extract catastrophe rents (Perry 2023). For example, coastal homeowners are only able to access home insurance because reinsurers can raise capital by selling storm damage risk to pension fund investors via catastrophe bonds (Taylor 2020). Thus, webs of international financial institutions, institutional investors, credit rating agencies, asset managers and bondholders 'connect the colonial plantation complex to ongoing forms of dependence and coercion' (Perry 2023: 567).

Perry's focus for ethical critique appears to be the wider social environment of the international sovereign debt markets. The 'conditional sovereignty' of these postcolonial states is premised on governing climate change through the 'racially hierarchical sovereign debt system' (Perry 2023: 567). The role of calculative practices – such as catastrophe risk modelling – is to 'normalize' debt as a response to the conditional sovereignty of Caribbean states and to repackage neocolonial burdens as fungible and tradable risks. The catastrophe risk industry effectively employs science to naturalize the uneven distribution of disaster vulnerability. The result is continued financial gain from geographical circumstances that are purportedly accidental.

Indeed, the international legal and financial system significantly constrains postcolonial sovereignty. Increasingly, monetary sovereignty is conceptualized as a spectrum that reflects the various degrees to which states can effectively exercise sovereign rights over money and finance (Bonizzi et al. 2019; Patrício Ferreira Lima 2022). The denomination of sovereign debt in US dollars links a debtor state's repayment capacity to the value of its GDP growth in US dollars. Therefore, the rate of its GDP growth will be more unstable, making it 'more difficult to ensure stable levels of debt sustainability despite fiscal planning through domestic law' (Patrício Ferreira Lima 2022: 434). Likewise, the distinct legal spatiality of sovereign debt conditions the national sovereignty of debt-issuing nations (Potts 2024). The absence of an international legal regime for sovereign insolvency resolution has created a fragmented and anarchic legal landscape (Nguyen 2013; Lienau 2016) in which sovereign debt litigation operates within national legislative frameworks. Approximately 50 per cent of all sovereign bonds are written under the law of the State of New York, with almost all remaining bonds written under the law covering the City of London. This is consequential as these legal systems empower private creditors over public debtors in cases of sovereign default and debt restructuring negotiations (Potts 2020). Since the 1990s, the traditional legal basis for sovereign states to discharge their debt obligations and suspend debt repayments, based on the public nature of the debt and their sovereign status, has been eroded in New York and English courtrooms (Sassen 2017).

Our contention is that in the case of sovereign debt disaster clauses, legal and monetary contexts are not the only conditions that constrain the sovereignty of disaster-vulnerable states. These clauses are activated once several economic and scientific facts are established. Our present objective is to provide a full account of how the design and execution of scientific and computational practices do more than simply legitimize unjust post-plantation power relations, showing how these technical practices actively constrain the sovereign decision-making space of disaster-vulnerable nations. In doing so, we offer an ethicopolitical critique of these non-human actors.

Louise Amoore's work (2020) on machine learning offers a starting point for this inquiry because it addresses a cognate methodological issue raised by the collective actions of humans and non-humans. Such phenomena, for example, driverless cars, can frustrate attempts at ethical critique because ascribing responsibility and intention to non-human actors appears counterintuitive. Amoore addresses the contemporary conceit that the algorithms of deep neural learning[3] 'render all agonistic political difficulties as tractable and resolvable'. In doing so, Amoore (2020: 12) argues that the 'single optimal output' or algorithmic 'truth claim' should be treated as a 'hinge point' for critical inquiry. This implies that the fact on which a claim rests is produced by 'multiple and contingent relations', which serve to overlook and conflate information that threatens to throw that claim into doubt (Amoore 2020: 12). As such, political and ethical interventions are made possible by algorithms which emerge within specific conditions. In the cases explored by Amoore (2020: 9), such 'conditions of emergence' include relations that are 'identifiably between humans and algorithms', such as 'the editing of code', alongside 'relations of algorithms to another algorithm, such as a classifier supplying the training data from which a neural network will learn'.

Treating disaster clauses as 'hinges' allows for the exploration of conditions under which states have the option to exercise a sovereign decision, including the suspension of debt repayments. The contractual clauses include parametric triggers which set standardized proxies to determine payouts and do so by simplifying, excluding and abstracting from contextual conditions (Johnson 2020). This process of setting proxies and loss estimates is simultaneously an exercise in ruling out some disasters as providing the conditions for the sovereign decision. Our attention to the design of clauses, and to the collection and analysis of the data that determines whether the clause has been activated, brings constraints on the sovereign decision to the forefront of the analysis. Disaster clauses therefore exemplify what Stuart Elden (2017: 300, 303) has called a 'regime of global calculation', in which the measurement of the material environment has emerged and operates as a constraint on modern statecraft.

Disaster Clauses in the Caribbean Region

In recent years, countries within the Caribbean region have incorporated climate-resilient debt clauses (CRDCs) into sovereign debt instruments, reflecting an evolving approach to managing the financial risks imposed by natural disasters. The intention behind the CRDCs is to allow recovering nations to focus their resources on emergency response and recovery rather than on debt repayments. This is to avoid leaving creditors and debtors in gridlock while the debtor focuses on the emergency. Once recovered from the initial economic shock of the disaster, these nations will resume repaying their creditors.

To date, three Caribbean states – Grenada, Barbados and the Bahamas – have implemented CRDCs. The CRDCs in Grenada have been evolving since 2005, following Hurricane Ivan in 2004. Initially a short-term solution, Grenada adjusted its interest rates and extended repayment deadlines to enable resource allocation for disaster relief. The year 2015 saw Grenada introduce a more advanced CRDC, allowing the nation to defer debt repayments if an eligible natural disaster that causes more than US$15 million in losses occurs. This iteration was more robust, embedding the CRDC within key financial agreements including US dollar-denominated bonds maturing in 2030, loans from the Export Bank of China and agreements with Paris Club creditors. Reporting and accountability mechanisms were also embedded, meaning Grenada would have to provide certification of trigger events, including written confirmation from its parametric insurance provider, the Caribbean Catastrophe Risk Insurance Facility (CCRIF).

Following Hurricane Beryl in 2024, further adjustments to the CRDC were made. Hurricane Beryl caused destruction across Grenada so severe that the nation received a US$42.42 million payout under the tropical cyclone policy with the CCRIF. Subsequently, Grenada will activate the Escape Clause under the Fiscal Resilience Act (FRA). Initially established in January 2024, the FRA is designed to maintain fiscal stability in extraordinary circumstances by setting fiscal targets for a government's budgetary performance. Under the Escape Clause of the FRA, Grenada will be able to bypass fiscal rules and debt-servicing obligations to allocate resources for disaster recovery. The Grenadian government has predicted the Escape Clause will continue to be activated throughout 2025 as the nation continues to rebuild following Hurricane Beryl.

Barbados's first iteration of its CRDC came in the early 2010s, enabling Barbados to defer debt repayments if an eligible natural disaster caused more than US$5 million in damage. An update to the CRDC came in 2019 to set new thresholds: US$5 million for earthquakes and floods and US$7.5 million for tropical cyclones. This update allows Barbados to free up US$700 million, or 15 per cent of its GDP, from debt-servicing obligations and towards resource

Table 5.1 An overview of sovereign debt disaster clauses for Grenada, Barbados and the Bahamas.

	Date issued	Type of CRDC	Bond currency	Trigger events	Key features	Legal jurisdiction
Grenada	2005	1st iteration of CRDC	US dollar	Hurricane Ivan (2004) and related natural disasters	Modified coupon structure and maturity dates	New York Law
	2015	Revised CRDC	US dollar	Tropical cyclones causing more than $15 million in loss and damage	Permits deferral of principal and interest payments based on the severity of the disaster; applicable up to three times	New York Law
	2024	Fiscal Resilience Act (FRA) activation. Applies to CRDCs in fiscal arrangements with IFOs	US dollar	Severe natural disasters such as Hurricane Beryl (2024)	Temporarily suspends fiscal rules to enable the allocation of resources for disaster recovery; permits deviation from the mandated primary balance of at least 1.5% of GDP under exceptional conditions	New York Law

Barbados	Early 2010s	1st iteration of the CRDC	US dollar	Tropical cyclone, earthquake and excess rainfall exceeding $5 million in loss and damage	Permits deferral of principal and interest payments for up to two years; payments adjusted pro rata; deferral possible up to three times	New York Law, London Law & Domestic Law
	2019	Revised CRDC	US dollar	Tropical cyclones with loss and damage exceeding $7.5 million; earthquakes and flooding with loss and damage exceeding $5 million	Deferral allowed up to three times but not in the final two years; includes a new 'bondholder veto' feature; potential to free up $700 million in debt-servicing obligations for disaster relief management	New York Law, London Law & Domestic Law
The Bahamas	2021	IDB loans: Contingent Loan Arrangement with the IDB containing a CRDC	US dollar	Eligible tropical cyclones with emergency declaration such as Hurricane Dorian (2019)	Permits deferral of principal repayments for up to two years; interest payments continue; does not extend loan maturity; applies to all eligible IDB loans	No national jurisdiction; disputes resolved through arbitration

Source: Authors' own, adapted from Fontana and Ho (2021); Private Sector Working Group [PSWG] (2021); IDB (2024) and Straker (2024).

allocation for disaster relief. Barbados has comprehensively embedded its CRDC across the entirety of public debt instruments, making its public debt stock the only climate-resilient debt stock globally. Greater global ambitions for the CRDCs have come from Barbados's Bridgetown Initiative, launched in 2022 by Prime Minister Mia Mottley. The Bridgetown Initiative aims to reform the global financial system to better support countries affected by natural disasters and climate change. This scheme has gained traction, with Western states such as Canada opting to include similar CRDCs in their sovereign lending.

The final Caribbean case is the Bahamas, where in 2021 a Contingent Loan Arrangement was activated with the Inter-American Development Bank (IDB) which contained a CRDC. Provoked by Hurricane Dorian in 2019, the Bahamas can defer repayments on its IDB loans for up to two years following an eligible natural disaster.

Producing Hinge Facts, Constraining Sovereignty

Our approach to disaster clauses scrutinizes the facts on which the clauses hinge, such as the cost of estimated losses, the type of trigger clauses, and the data employed to measure the natural disaster and estimate losses relating to it. In what follows, we show how these facts are established and the ways in which these processes constrain or preclude the sovereign decision of three Caribbean governments.

Parametric Specification of Clause Triggers

While the legal and monetary regimes governing disaster clauses are uniform, there is more variation in the facts about weather and economic loss on which disaster clauses hinge. The disaster clauses negotiated by the Bahamas are premised on the activation of parametric triggers. These parametric techniques for loss calculation assume 'a scalar correspondence between selected variables at easily observed scales, and actual losses at less easily observed scales' (Johnson 2020: 48). Such triggers are pre-established threshold parameters that are quantifiable features of natural hazards such as hurricane wind speed (Schuster 2021). The disaster clause would therefore be triggered if the event exceeds pre-established parameters (Johnson 2021).

Alternatively, both Grenada's and Barbados's disaster clauses employ parametric index triggers. These triggers hinge on a higher burden of proof because payments are premised on both the event's intensity and the scale of losses incurred, as determined by catastrophe modelling software (Commonwealth Secretariat 2016). This makes a debtor's decision to suspend debt repayments conditional on two low-probability events being modelled as occurring concurrently. For example, a strong Category 4 storm would trigger a claim only if it caused damage above a specified threshold. This difference could be down to the same storm damaging some valuable buildings and not less valuable ones nearby. Given these fine spatial margins, parametric index triggers provide a narrow corridor for sovereign decisions.

The Collection, Analysis and Funding of Data

The CCRIF is a regional risk pool which issues parametric insurance across nineteen Caribbean governments and eleven Central American governments,[4] offering earthquake, tropical cyclone and excess rainfall policies. The primary purpose of the CCRIF is to provide short-term liquidity to members that experience predefined disasters. The pooling of risks into a single diversified portfolio lowers the insurance capital needed for payouts and reduces the insurance costs for members. Payouts from parametric insurance are based on impact and loss estimated within a catastrophe risk model, making them a cost-effective approach to insurance, as the need for a post-disaster loss assessment is removed.

CCRIF datasets and infrastructures employed in the gathering and computation of that data have been reharnessed within sovereign debt disaster clauses, allowing for the creation of a mechanism to normalize relations between sovereign bondholders and their disaster-vulnerable debtors. The scope for bond issuers to suspend repayments depends on the collection and analysis of data and the acceptance of the methods, datasets and infrastructures employed in the gathering and computation of that data. Consequently, elements of the CCRIF provide conditions in which sovereign decisions emerge.

The inclusion of the hurricane provision in each restructuring agreement depends on creditors regarding the data underpinning the assessment of risks as legitimate (Commonwealth Secretariat 2016). Creditors must also submit to the independence and expertise involved in the assessment of catastrophe risks and damage incurred. Disaster clauses – and the conditions for constrained sovereign decisions – thus elicit confidence from creditors when they have leveraged desirable traits of 'good' scientific research such as claims of objectivity and an impartial analytical approach.

The sources of data which assured creditors that the information informing the analysis was impartial, robust and objective have been historically implicated in both the development of US military capability and what Sara Grossman (2023: 3) has labelled settler-colonial 'world-building'.[5] The weather data had two main sources, namely, the US-based National Hurricane Center (NHC) and the CCRIF itself. The NHC obtains its data from the National Oceanic and Atmospheric Administration (NOAA). Formed in the 1970s, NOAA was incorporated into the US Department of Commerce. NOAA builds on the work of four older federal agencies: the Weather Bureau of the United States, the Bureau of Commercial Fisheries, the Coast and Geodetic Survey Corps and the United States Coast and Geodetic Survey. The satellite meteorology used to gather much of NOAA's data was developed with support from the US Department of Defense and the Central Intelligence Agency.

We find similar legacies of American militarization and geostrategic interests in several of the datasets and analytical systems that the CCRIF employs. The CCRIF had already collected an extensive dataset of Caribbean natural disasters and undertaken modelling work on it, including substantial surveillance of each participating member of the insurance pool. For instance, to provide modelled losses for disasters, the CCRIF uses population data from Landscan. The Landscan database analyses census and remote sensing data through dissymmetric mapping, which is a geospatial technique that mobilizes information about land use and cover, slope and proximity to infrastructure and distributes population counts within selected boundary areas. Landscan is typically used for risk planning purposes in the contexts of humanitarian response, disease mapping, risk analysis and evacuation modelling. This dataset is produced by Oak Ridge National Laboratory in Tennessee and administered by UT-Battelle on behalf of the U.S. Department of Energy.

The CCRIF's real-time forecasting system also draws on data from the Automated Tropical Cyclone Forecasting System (ATCFS). The ATCFS comprises a set of tools that graphically display cyclone tracks, analyse information and generate warning messages and is another database intimately connected to the military capability of the United States. It was developed by the U.S. Naval Research Laboratory (NRL) in Monterey, California, which is the Marine Meteorology Division of NRL's Ocean and Atmospheric Science and Technology Directorate (Roy and Kovordányi 2012: 63). This part of the NRL undertakes research that captures how fleet operations are affected by atmospheric processes and develops 'automated systems that analyse, simulate and predict . . . the structure and behaviour of these processes and their effect on naval weapons systems' (NRL, n.d.). Ultimately, this research and forecasting underpin the US Navy's worldwide environmental forecasting capability.

As the emergence of disaster clauses in the sovereign debt restructuring agreements of some CCRIF members illustrates, both the NOAA and Landscan datasets are considered sources of independent, impartial and objective data. However, these datasets are far from depoliticized because they are funded by various components of the US federal government. Given the significance placed on this independent and objective data by creditors, disaster clauses risk becoming dependent on the continued funding of NOAA and Landscan by the US government. The wider implication of this political economy is that the sovereign decision to suspend debt repayments is subject to the vicissitudes of US domestic politics and the political cycles of another sovereign state.

Geometric Constraints on the Sovereign Decision

An important corollary to the funding and ownership of the data on which disaster clauses are premised is the scope and fungibility of the CCRIF, Landscan and NOAA data that is modellable. The type and quality of data constitutes a second limit on the sovereign decision, because it helps to determine the types of disaster that can be subject to debt suspension clauses. The CCRIF's approach to parametric insurance models has evolved since its inception in 2007. Initially, the CCRIF used proprietary models from the risk modelling vendor EQECAT for hurricane and earthquake hazards. However, due to data gaps and limitations, the second-generation modelling framework, known as the hazard and loss estimation model (HLEM), was established by the CCRIF and partners. The HLEM would be succeeded by the multi-peril risk estimation system (MPRES) in 2011.

As member governments sought additional coverage, the CCRIF introduced an excess rainfall (XSR) product in the 2012/13 policy year, developed with Swiss Re and Kinetic Analysis Corporation using NASA satellite data. This product has been updated several times, the latest version being the ESR2.5 model.

Starting from the 2019/20 policy year, the CCRIF replaced the MPRES model with its own SPHERA models for tropical cyclones and earthquakes. The SPHERA models are designed to enhance precision, offer detailed estimates of losses and use high-resolution data from the Landscan database. This model incorporates historical loss data from past events in the Caribbean and Central America.

Data Silences

The latest versions of SPHERA and ESR2.5 model weather and physical data relating to wind speed, wind and storm surge, land and underwater topography, seismic waves, ground motion, temperature, pressure, humidity and precipitation. At present, these models do not cover the impacts of landslides, despite this being a significant peril in the Caribbean due to the region's geology and climate (DeGraff et al. 1989). This type of disaster is also deeply intertwined with place-specific considerations of environmental and climate justice. Barbados was the first English colony to commercially cultivate sugar cane from 1627. This period, known as the 'Sugar Revolution', oversaw a near-total transformation of the Barbadian economy, infrastructure and natural landscape for the purpose of imperial sugar cane exports (UNESCO 2014). To accommodate the extensive sugar cane exports, Barbados's tropical forests were cut back and cleared (Hollsten 2008; Webber 2018). However, in tropical environments such as Barbados, a significant proportion of the ecosystem's nutrients is stored in vegetation rather than in the soil, meaning extensive deforestation leaches the soil of nutrients. Through removing deep-rooted trees from an area, the land becomes increasingly unstable. This, coupled with a system of monoculture plantations where sugar cane was replanted annually, culminated in severe soil degradation (Hollsten 2008). Relentless exploitation of Barbadian soil under colonialism has left the nation vulnerable to landslides, exacerbated by the island's naturally hilly geography and heavy seasonal rainfall, which further destabilize the soil.

Similarly, following the handover from French to British colonial rule in the eighteenth century, Grenada became a hub for colonial agricultural exploitation, exporting cash crops such as sugar, coffee and cocoa. Using slash-and-burn agricultural methods, Grenada's vegetation and forested areas were destroyed to make way for monoculture plantations (Ternan et al. 1989). This aggressive exploitation has led to nutrient-deprived, deteriorated soil, harming land stability in Grenada. The topography of Grenada, being hilly and mountainous, is already naturally prone to landslides. A combination of natural risk and colonial exploitation leaves Grenada highly vulnerable to landslides (Van Westen 2016). Despite a well-documented history of this vulnerability, the CCRIF does not provide landslide coverage within parametric policies due to gaps in the available data. As such, states continue to be constrained in decision-making due to vulnerabilities stemming from colonial injustices.

Scalar Obstacles

Our work on the politics of calculation demonstrates that sovereignty is not merely constrained by the absence of data. Indeed, a second factor constraining sovereignty is the limitations of the modelling that is facilitated by the CCRIF, NOAA and Landscan datasets. For example, the inclusion of data, modelling and coverage for rainfall would introduce a significant driver of basis risk into disaster clause coverage. This refers to the risk that the conditions estimated by the parametric measure diverge from those that the policy buyer experiences on the ground. In this instance, the money that the insured customer may receive as a payout could be significantly higher or lower than the full cost of the trigger event. While basis risk affects all forms of insurance, parametric insurance is particularly vulnerable to it due to the series of scalar abstractions made in the process of contract design. Such geographical assumptions are attempts by contract designers to align observed variables at one scale with actual losses at another (Johnson 2021). For example, 'spatial basis risk' occurs when contract designers need to find an observable spatial unit small enough to make the abstraction meaningful yet large enough to facilitate straightforward sales administration. This type of basis risk is especially problematic for areas with high rainfall variability and microclimates.

There are good reasons to think that some disasters are problematic proxies to include in sovereign debt disaster clauses. As Leigh Johnson (2021) has noted, basis risk frequently occurs in parametric insurance contracts relating to rainfall, and insurers have responded to this problem by offering insured parties discretionary payouts. In such a way, problems with parametric calculations have been resolved using an insurance practice historically employed by insurers to avoid disputes over contractual issues and customer discontent: the 'ex gratia' payment. The CCRIF has made discretionary ex gratia payouts to participants in its rainfall coverage pool, the XSR SP. For example, in the insurance year 2022/23, two events triggered claim payments amounting to US$6,263,359, while two participating countries received ex gratia payments totalling US$1,855,000.

Even though a number of Caribbean countries are vulnerable to excess rainfall, this peril is a problematic trigger for sovereign debt disaster clauses. The calculative errors that motivate ex gratia payments effectively 'reopen' the sorts of 'negotiations and contestations' (Johnson 2021: 129). This would undermine the rationale for including disaster clauses in debt contracts, which was to avoid further restructuring negotiations. Moreover, even if rainfall were to be included in such disaster clauses, it seems likely that this hazard would reproduce the power relations embedded within post-plantation financial ecologies. In the realm of weather insurance, negotiations are undertaken 'on insurer's terms … thus informalising the power relations embedded in parametric insurance' (Johnson 2021: 129). In the realm of sovereign debt restructuring, we might expect basis risk and negotiations over the activation of clauses to reintroduce debtor–creditor power relations and the circumscribed legal status of sovereign states. Subsequently, the errors in modelling some disasters may only be resolved through informal mechanisms that constrain the sovereign decision. Therefore, the basis risk attached to some disasters draws our attention to the relationship between calculation and the channels through which contemporary global governance plays out (Elden 2017).

Conclusion

To date, geographical scholarship on disaster management has employed relational thought to account for the way disasters are managed through networks of diverse entities, which often work together despite being individually located in separate places. This chapter has explored how scientific and financial calculative practices constrain the sovereign decision-making space of disaster-vulnerable nations, while also offering an ethical critique of these non-human actors. To accomplish this, we have scrutinized disaster clauses in sovereign debt contracts. These clauses in sovereign bonds give a country's government the option to temporarily suspend its debt burden should it experience weather or climate disasters that meet predefined parameters. Such clauses require quantifiable and verifiable indicators of financial losses and/or disaster magnitude. By treating the seemingly technical and scientific facts on which these clauses rest as 'hinges', this chapter traces the 'multiple and contingent relations' which serve to exclude information in ways that foreclose a country's sovereign decision to suspend debt repayments (Amoore 2020).

The hinge points in the disaster clauses of debt issued by Grenada, Barbados and the Bahamas support three observations about the constrained sovereignty of disaster-vulnerable states. First, the type of trigger for the clause greatly influences the likelihood of national governments being presented with the decision to suspend debt repayments. If the clause contains a parametric trigger based on the geophysical properties of the relevant disaster, then the debtor will be presented with the sovereign decision regardless of where the event impacts within its territory or the damage it causes. Alternatively, parametric index triggers further constrain the scope for the sovereign decision because the clause will be activated by both the properties of the disaster itself and the specific economic impact of the event. This renders the sovereign decision more susceptible to the contingencies of geography and therefore shrinks the space and time of that decision.

Second, creditors agree to disaster clauses when they have confidence in the objectivity and impartiality of the data used to model losses and the measurement of disasters. In the cases we have studied, the relevant datasets have been collected using scientific infrastructures and techniques deeply rooted in settler-colonial histories and have been funded by the US federal government in the pursuit of militarization and energy security. The latter finding highlights that the sovereign decisions of Grenada, Barbados and the Bahamas are subject to continued funding by the US government and vulnerable to political decision-making beyond their control.

Third and finally, the datasets themselves impose constraints on sovereign decisions by virtue of the geophysical phenomena that can be measured and the limitations of data that are measured. The data and models employed within the disaster clauses we studied do not produce loss estimates for the impact of landslides. Not only does this mean that the sovereign decision is not available for the Caribbean's most commonly occurring disaster, but it is a disaster vulnerability that is directly linked to land use and economic activities in the plantation system. Other types of data relating to excess rainfall are vulnerable to basis risk, which can undermine the activation of disaster clauses and reintroduce negotiations between creditors and debtors through informal channels that diminish the bargaining power of sovereign states.

Notes

1 This category is applied to hurricanes with a wind speed of 156 miles per hour, causing destruction of buildings, trees and signs. Such hurricanes require evacuation of up to 10 miles inland (Met Office, n.d.).

2 Derived from the historical experience of plantation slavery, this term is used to describe a 'system of power' in which marginalized and racialized 'subjects labour for the benefit of others' (Clukey and Wells 2016: 2). In such a way, it provides a 'theoretical framework for thinking about the ways Black life and Black histories link to post-slavery conceptualizations of geographic violence' (McKittrick 2013: 4).

3 This is a method in artificial intelligence that trains computers to process data in a way that is loosely modelled on the structure and function of the human brain. The aim is for the computer to recognize complex patterns in data to generate accurate insights and predictions.

4 As an insurance mechanism it has paid out US$265 to seventeen member governments up to 2023.

5 For example, Grossman (2023: 151) highlights how Landsat satellites were used to develop conceptual site plans and conduct habitat and impact assessments for oil pipelines that cross indigenous territories and undermine land sovereignty.

References

Amoore, L. (2020), *Cloud Ethics: Algorithms and the Attributes of Ourselves and Others*, Duke University Press.

Anderson, B., M. Kearnes, C. McFarlane and D. Swanton (2012), 'On Assemblages and Geography', *Dialogues in Human Geography* 2 (2): 171–89.

Bonizzi, B., A. Kaltenbrunner and J. Michell (2019), 'Monetary Sovereignty Is a Spectrum: Modern Monetary Theory and Developing Countries', *Real-World Economics Review* 89:46–61.

Clukey, A. and J. Wells (2016), 'Introduction: Plantation Modernity', *Global South* 10 (2): 1–10.

Commonwealth Secretariat (2016), 'Introducing Hurricane Clauses: Lessons from Grenada's Recent Experience'. Available at: https://production-new-commonwealth-files.s3.eu-west-2.amazonaws.com/migrated/inline/Introducing%20Hurricane%20Clauses.PDF (accessed 7 November 2025).

Cotterill, J. and L. Harris (2024), 'Grenada Triggers "Hurricane Clause" to Suspend Bond Payments', *Financial Times*, 23 August. Available at: https://www.ft.com/content/06bdabb2-2abb-45ab-9ee4-94e1c328598f#:~:text=Grenada%20told%20investors%20in%20its%20$112mn%20bond,year%20as%20it%20recovers%20from%20Hurricane%20Beryl (accessed 7 November 2025).

DeGraff, J. V., R. Bryce, R. W. Jibson, S. Mora and C. T. Rogers (1989), 'Landslides: Their Extent and Significance in the Caribbean', *Landslides: Extent and Economic Significance* 68:51–80.

Donovan, A. (2017), 'Geopower: Reflections on the Critical Geography of Disasters', *Progress in Human Geography* 41 (1): 44–67.

Duncan, N. (2024), 'Caribbean Leaders Call for "Marshall Plan" to Help Rebuild After Hurricane Beryl', *Guardian*, 18 July. Available at: https://www.theguardian.com/world/article/2024/jul/18/caribbean-leaders-marshall-plan-rebuild-hurricane-beryl (accessed 7 November 2025).

Elden, S. (2017), 'Foucault and Geometrics', in P. Bonditti, D. Bigo and F Gros (eds), *Foucault and the Modern International: Silences and Legacies for the Study of World Politics*, 295–311, Palgrave Macmillan US.

Fontana, S. and S.-J. Ho (2021), 'Sovereign Debt Evolution: The Natural Disaster Clause', *Emerging Markets Restructuring Journal* 11:1–14.

Grossman, S. J. (2023), *Immeasurable Weather: Meteorological Data and Settler Colonialism from 1820 to Hurricane Sandy*, Duke University Press.

Grove, K. (2013), 'From Emergency Management to Managing Emergence: A Genealogy of Disaster Management in Jamaica', *Annals of the Association of American Geographers* 103 (3): 570–88.

Grove, K. (2014a), 'Agency, Affect, and the Immunological Politics of Disaster Resilience', *Environment and Planning D: Society and Space* 32 (2): 240–56.

Grove, K. (2014b), 'Biopolitics and Adaptation: Governing Socio-ecological Contingency Through Climate Change and Disaster Studies', *Geography Compass* 8 (3): 198–210.

Grove, K. and J. Pugh (2015), 'Assemblage Thinking and Participatory Development: Potentiality, Ethics, Biopolitics', *Geography Compass* 9 (1): 1–13.

Hollsten, L. (2008), 'Controlling Nature and Transforming Landscapes in the Early Modern Caribbean', *Global Environment* 1 (1): 80–113.

Inter-American Development Bank (IDB) (2024), 'Disaster Risk Financing Instruments'. Available at: https://www.ccrif.org/sites/default/files/2024-cari-reg-workshop/Day3/Day3-Disaster-Risk-Financing-IDB.pdf (accessed 7 November 2025).

Jarzabkowski, P., K. Chalkias, E. Cacciatori and R. Bednarek (2023), *Disaster Insurance Reimagined: Protection in a Time of Increasing Risk*, Oxford University Press.

Johnson, L. (2020), 'Sharing Risks or Proliferating Uncertainties?: Insurance, Disaster and Development', in I. Scoones and A. Stirling (eds), *The Politics of Uncertainty: Challenges of Transformation*, 45–57, Routledge.

Johnson, L. (2021), 'Paying Ex Gratia: Parametric Insurance after Calculative Devices Fail', *Geoforum* 125:120–31.

Leyshon, A. (2020), 'Financial Ecosystems and Ecologies', in J. Knox-Hayes and D. Wójcik (eds), *The Routledge Handbook of Financial Geography*, 122–41, Routledge.

Lienau, O. (2016), 'The Challenge of Legitimacy in Sovereign Debt Restructuring', *Harvard International Law Journal* 57:151.

McGowran, P. (2024a), 'Geopolitical Assemblages and Disasters-in-the-Making in Kalimpong, India', *Political Geography* 108:103008.

McGowran, P. (2024b), 'Landslide Disasters in Kalimpong, India: Matters of Time?', *Geoforum* 156:104140.

McGowran, P. and A. Donovan (2021), 'Assemblage Theory and Disaster Risk Management', *Progress in Human Geography* 45 (6): 1601–24.

McGowran, P., H. Johns, E. Raju and S. Ayeb-Karlsson (2023), 'The Making of India's COVID-19 Disaster: A Disaster Risk Management (DRM) Assemblage Analysis', *International Journal of Disaster Risk Reduction* 93:103797.

McKittrick, K. (2013), 'Plantation Futures', *Small Axe: A Caribbean Journal of Criticism* 17 (3 (42)): 1–15.

Met Office (n.d.), 'Measuring Tropical Cyclones'. Available at: https://www.metoffice.gov.uk/weather/learn-about/weather/types-of-weather/hurricanes/measuring (accessed 7 November 2025).

Nguyen, D. V. (2013), 'Too Big to Fail – Towards a Sovereign Bankruptcy Regime', *Cornell International Law Journal* 45:697.

Patrício Ferreira Lima, K. (2022), 'Sovereign Solvency as Monetary Power', *Journal of International Economic Law* 25 (3): 424–46.

Perry, K. K. (2022), 'Ungovernable Islands, Burdensome Debts and Racialized "Disaster" Zones: The Caribbean as the Last Frontier of Capitalist-Colonial – Climate Crises'. Available at: http://dx.doi.org/10.2139/ssrn.4095348.

Perry, K. K. (2023), 'From the Plantation to the Deep Blue Sea: Naturalising Debt, Ordinary Disasters, and Postplantation Ecologies in the Caribbean', *The Geographical Journal* 189 (4): 562–74.

Perry, K. K. (2024), 'Epistemic Silences, Subversive Politics: Post-disaster Economic Assessments as Technologies of Persistent Coloniality and Route to an Emancipatory Climate Justice Agenda in the Caribbean', *Climate and Development* 16 (9): 798–810.

Potts, S. (2020), '(Re-)writing Markets: Law and Contested Payment Geographies', *Environment and Planning A: Economy and Space* 52 (1): 46–65.

Potts, S. (2024), *Judicial Territory: Law, Capital, and the Expansion of American Empire*, Duke University Press.

Private Sector Working Group (PSWG) (2021), 'Climate Resilient Debt Clauses (CRDCs)'. Available at: https://www.icmagroup.org/assets/Chairs-Summary-UK-Chaired-Private-Sector-Working-Group-Sub-Group-on-CRDCs.pdf (accessed 2 September 2025).

Roy, C. and R. Kovordányi (2012), 'Tropical Cyclone Track Forecasting Techniques – A Review', *Atmospheric Research* 104:40–69.

Sassen, S. (2017), 'Embedded Borderings: Making New Geographies of Centrality', *Territory, Politics, Governance* 6 (1): 5–15. https://doi.org/10.1080/21622671.2017.1290546.

Schuster, C. E. (2021), 'Weedy Finance: Weather Insurance and Parametric Life on Unstable Grounds', *Cultural Anthropology* 36 (4): 589–617.

Sheller, M. (2018), 'Caribbean Futures in the Offshore Anthropocene: Debt, Disaster, and Duration', *Environment and Planning D: Society and Space* 36 (6): 971–86.

Straker, L. (2024), 'Government to Trigger FRA Escape Clause Following Hurricane Beryl', *Now Grenada*. Available at: https://nowgrenada.com/2024/07/government-to-trigger-fra-escape-clause-following-hurricane-beryl/ (accessed 9 September 2025).

Taylor, Z. J. (2020), 'The Real Estate Risk Fix: Residential Insurance-Linked Securitization in the Florida Metropolis', *Environment and Planning A: Economy and Space* 52 (6): 1131–49.

Ternan, J. L., A. G. Williams and C. Francis (1989), 'Land Capability Classification in Grenada, West Indies', *Mountain Research and Development* 9 (1): 71–82.

United Nations Educational, Scientific and Cultural Organization (UNESCO) (2014), 'The Industrial Heritage of Barbados: The Story of Sugar and Rum', UNESCO World Heritage Convention. Available at: https://whc.unesco.org/en/tentativelists/5942 (accessed 7 November 2025).

US Naval Research Laboratory (NRL) (n.d.), 'Marine Meteorology'. Available at: https://www.nrl.navy.mil/marine-meteorology/ (accessed 7 November 2025).

Van Westen, C. K. (2016), 'National Scale Landslide Susceptibility Assessment for Grenada'. Available at: https://www.researchgate.net/publication/305115056_National_scale_landslide_susceptibility_assessment_for_Grenada (accessed 7 November 2025).

Waterton, E. and J. Dittmer (2014), 'The Museum as Assemblage: Bringing Forth Affect at the Australian War Memorial', *Museum Management and Curatorship* 29 (2): 122–39.

Webber, O. (2018), 'The Plantation's Role in Enhancing Hurricane Vulnerability in the Nineteenth-Century British Caribbean', *Alternautus* 5 (2): 29–42.

6

Plants for Recovery: Indigenous Women's Perspectives in a Post-disaster Resettlement

An Approach from Feminist and Post-disaster Geographies

Ana Julia Cabrera Pacheco, Diego Antonio Reanda Sapalú, Lisa Mackenzie and Teresa Armijos Burneo

We arrived at Sonia's house early in the morning. She walked us through her home garden where she had trees, flowers, a few chickens and plants growing in pots hanging by the windows. She told us that her son was given some seeds in school to learn about the cycle of plants, and she took the opportunity to grow them in plastic bottles which hung around her garden. 'This is good', she told us, 'but what I really want is to be able to use the little land around my house to plant them directly into the soil'.

This chapter explores the significance and benefits of agri-food traditions in post-disaster contexts. We do so by focusing on Maya Tz'utujil women who were displaced after a large landslide that destroyed the community of Panabaj on the shores of Lake Atitlán, Guatemala, in 2005. The women use Maya knowledge about food and medicinal plant cultivation, passed down through generations, and share it with one another as they tend any available growing spaces around their homes in their new settlement. The significance and benefits of traditional agri-food knowledge systems in post-disaster localities are understood as crucial to formulating approaches to disaster response that are more context and culturally specific. Although the chapter touches upon the potential harms that being separated from *home* can have on both the physical and mental health of communities, it seeks to demonstrate how the Tz'utujil women cultivate their new landscape, working to continue life and engage in transformative action in the aftermath of disaster.

Sonia is part of a small, organized group of women, *Ixquii' Ajtz'utjilaa'* (Tz'utujil women, in the Mayan Tz'utujil language), who live in Chuk Muk, a post-disaster resettlement town located in Santiago Atitlán, Guatemala. It was built to accommodate approximately 900 displaced Tz'utujil families from Panabaj village. These families were resettled between 2008 and 2011, a few kilometres away from their old homes. Chuk Muk's planned and ultimately built environment did not support the Maya Tz'utujil families' needs to attend to and care for their own land – *attend* both in terms of practical interactions with the land in the form of cultivation and in supporting families to establish psychological and social connections between the Tz'utujil people and the new space they came to inhabit. In addition, the Chuk Muk settlement is surrounded by privately owned land which the Tz'utujil community cannot access for foraging or to cross to reach the shore of Lake Atitlán, just 700 metres away. However, growing their own food is an essential feature of the Tz'utujil families' existence because their earnings are hardly enough to maintain their food, transportation, education and health. Consequently, some families returned to Panabaj, hoping to acquire pieces of land, given that it is one of the few places where land remains affordable in the region, or they return temporarily to the land that they still retain (Cabrera Pacheco et al. 2024).

The deterioration and loss of traditional socio-cultural practices after a catastrophic disaster and the specific ties of those practices to people, land and place are commonly overlooked in the design and delivery of post-disaster policies and governmental actions. Agri-food traditions exemplify the critical and often most tangible relationships between people, their homes, their livelihoods and their land. It is from this vital context in a Maya locality that we seek to learn. In Guatemala and in the village of Chuk Muk, Maya ways of tending to the land offer a counterbalance to the severing forces of displacement. That is, gentle, correlative practices of growth in a new territory generate small-scale but discernibly active and process-based relationships between women and land that lead to new forms of inclusivity *with* place.

We draw on (post)disaster geographies and feminist geographies to guide and inform our work. Post-disaster geographies employ interdisciplinary methodologies to understand the implications of disasters upon social life and the multifaceted and multilayered manifestations in place, space and time of disasters. Feminist geographies give a framework to link everyday lives and care practices of bodies and territories and to situate centrally the relevance of gender-specific experiences during disasters and the long-term processes that follow. Methodologically, we followed a feminist decolonial approach (Curiel 2021; Armijos and Ramírez Loaiza 2024) that pushes for accountability in research while engaging with diverse knowledges that come together in dialogue. It brings forward situated experiences and other ways of being in the world through memories, traditions and stories (Naylor et al. 2018; Kovach 2021; Zaragocin and Caretta 2021; Smith 2022). Our research was developed as a participatory, creative and collaborative process where the main objective was to explore how the agri-food traditions of this group of women make possible the continuous reproduction of Maya ways of life in post-disaster contexts. Our empirical findings are set in two sections, where we differentiate between scales of analysis, the home garden as a site for recovery, and the bodies that inhabit them which are under recovery processes through their relation to plants.

In summary, this chapter argues that growing food and harvesting plants using traditional knowledge helps displaced communities, and specifically indigenous women, to gain a positive sense of connection and connectivity back into the territory in which they now live their lives.

(Post)disaster Geographies: Recovery and Resettlement

A key agenda of (post)disaster geographies is to understand the politics and challenges of place-making after a disaster. Scholarship in post-disaster geographies reflects on past events or ongoing disasters to understand the land and livelihood processes of re-territorialization, working to reveal the unique forms of resistance and solidarity that can materialize in the aftermath of a destructive event (Sou 2022). Although a disaster may follow an event of natural origin, it is the combining forces of vulnerability, exposure and recovery capacity within the affected community that so often define the extent and time frame of devastation upon people. In many cases, these forces were constructed systematically long before the 'event', often beyond the control of people. In Panabaj, for example, in the 1980s, the Guatemalan army established a military post during the armed conflict,[1] which forced people to be displaced to some of the areas that were eventually buried in the 2005 landslide. As we can see here, people in areas exposed to high environmental risk, where land is often cheap or undesirable, are usually those already dealing with precarity or poverty (Vergara-Saavedra 2024).

The social production of disaster risk is further mediated by hierarchies of race, class and gender which deeply determine individuals' vulnerability to disasters (Bonilla 2020; Armijos Burneo and Ramírez Loaiza 2021; Vergara-Saavedra 2024). Feminist approaches have also uncovered how structural violence(s) are exacerbated during disasters, where in the fight for survival and the continuation of life, women's rights and well-being lose ground. For example, Bradshaw and Fordham (2015: 238) address 'the escalation of hours in the working day' for women and girls but also the increased risks of permanently interrupting the education of girls. Thus, disasters are historically built upon the bodies of women and nature (González-Ramírez et al. 2022). Indeed, these violent processes do not go away once the hazard 'event' has passed but remain in people's memories and in their bodies, as well as within their territories (Adams-Hutcheson and Smith 2020). Disasters are therefore life-changing processes for the affected people, and their impacts are far-reaching (Watts and Brannum 2023). They occur across plural ways of living, which have not yet been fully considered in disaster and risk management studies, particularly those located in indigenous communities. A generalization of the need for resilience fails to uncover the full and complex socio-environmental conditions that communities face, which are both place and historically specific.

Resettlements are usually a generalized 'consequence' of disasters to which affected communities must simply 'adapt'. Resettlements are projected and constructed by external 'experts', usually without any input or participation from the displaced communities. As such, plans and designs may undermine communities' ways of life, including their everyday social reproduction (Fernández et al. 2019). For this reason, Taiban et al. (2020: 12) argue that post-disaster policies 'must be rooted in the daily habits of the residents'. Resettlements must be designed and delivered in ways that work with and not against the everyday lives, needs and concerns of disaster-affected peoples. This is particularly true when the affected communities depend on the land for their livelihoods. Indeed, planning decisions that are sensitive to the meaning of rootedness in relation to land-based practices would allow communities to better adjust and thrive in their new homes following displacement (Taiban et al. 2020). Thus, in following a (post)disaster geographies epistemology, this body of work seeks to counter political decisions, made by those in positions of power, taken without dialogue, by which people lose control of their own sense of belonging to place and their own destiny.

We use feminist geographies as well as creative and participatory methodologies to ground our work because they allow us to explore and understand the knowledges, memories and plants of indigenous women living in a post-disaster recovery context. These feminist frameworks allow us to unearth and understand indigenous women's knowledge and practices of inhabiting their new 'resettled' territory and exemplify alternatives to 'development' discourses where collective well-being and modest but significant ways of 'good living' manifest in culturally specific ways (Armijos Burneo and Ramírez Loaiza 2021). Through learning from conversations with the women about the contexts of both Panabaj and Chuk Muk, the very personal experiences of building a sense of home and belonging have been brought to the fore. We have also come to understand the way in which very specific practices of growing have led to a complex form of becoming *with* the territory through working with seeds and soil of the ground. Our research reflects on these contingent practices to consider ways to bridge the gap between inclusion and exclusion of their ways of life as a means for disaster recovery.

Feminist Geographies: Bodies and Territories

By drawing on feminist geographies, we acknowledge women's different positionality in the social life of the world and the way in which women experience geographical scales, connectivities to place and power, and hierarchies differently. By following feminist geographies, we seek dynamic, cooperative and connected practices with space, place and nature that are embodied. Such an approach requires sensitivity and engagement in our research practices as well as a focus on an individual scale. By centring more intimate scales of enquiry within resettled communities, we have been able to follow the women's contributions to the 'ongoingness' of their world, tracing their recovery practices in the very sites, conditions and matter of their activities (Massey 2013; Haraway 2016: 132). Thus, our work in understanding both their collective and personal acts of shaping and appropriating new ground in the territory is situated and seeks to draw from their experiences of relating to their surroundings.

Feminist geographies help us to situate women's bodies within their respective territories, but they also guide our understanding of how women construct their territories from their bodies and experiences (Naylor et al. 2018). For indigenous scholars Lorena Cabnal and Mishuana Goeman, 'the body is both a site and source of memory, knowledge, healing, and resistance, connected to humans and nonhumans through [multiple] scales of connectivity' (Vasudevan et al. 2023: 1738). Everyday embodied practices support the continuity of life within indigenous territories (Truelove 2011; Ulloa 2021). Particularly at the micro-scale, home gardens link 'contexts, necessities, and projects of Maya individuals, families, and communities', articulating their present and their history (Velázquez Solís 2012 in Cabrera Pacheco 2024: 10). Home gardens become integral spaces for community well-being based on self-provision with multiple livelihood activities such as food crops, medicinal herbs, animal husbandry, cooking and weaving, among others, which are well suited to the territory (Camacho Benavides 2020; Cabrera Pacheco 2022, 2024). The everyday actions these women perform to support social reproduction and maintain their cultural practices, including cultivating and caring for plants, are small but significant acts of resistance. In this sense, Sou (2022: 21) has analysed home gardens in post-disaster contexts as sites of everyday resistance. She shows how in Puerto Rico after Hurricane Maria, 'the women's garden represents a creative grassroots strategy to adapt to, mitigate and recover from the impacts of the hurricane'. If interpreted only as a resilient process, 'the political significance of the women's initiative will be missed' (Sou 2022: 21). This is not to say that women framed their garden as political or an act of resistance, but their purpose in leveraging multiple forms of response and recovery is certainly clear.

It is important to state that the connection between bodies, communities and territories can also manifest in negative forces such as risk and disease. To protect against future harm from disease, at this micro-scale, food and medicines are grown and rituals are undertaken. In these ways, women collectively care for their bodies, as healing the body can also mean healing the land in some indigenous contexts (Armijos Burneo and Ramírez Loaiza 2021) and vice versa (Kimmerer 2020). As Gladys Tzul Tzul (2015: 93), a Maya K'iche' activist, reflects, the territory is 'the material base for organising life'. Through these practices, women defend the connections between bodies and territories, where nature intersects to construct the socio-cultural world in its present form. The relational capacities of the Tz'utujil women and the affective and indirect forces that they experience carry forward into the ways in which they live their lives with the territory, taking steps forward, improving, refining, forming contingencies and shaping new lives.

Embodied Knowledges and Conversations from the Plants

Based on a feminist decolonial approach, this research is conceived as a process, and as such, it is co-constructed by all the co-investigators: researchers and the *Ixquii' Ajtz'utjilaa'*, a group of Tz'utujil women formed during this research.

Our research approach means to learn, unlearn and relearn through multiple conversations, walks, drawings and practice (Kovach 2021; Smith 2022; Walsh 2023; Armijos and Ramírez Loaiza 2024). To 'think from and with . . . knowledges grounded in territory and place' (Walsh 2023: 114). We, therefore, aimed to come together in dialogue from our diverse situated knowledges and experiences to understand the ways of being in the world of this group of Tz'utujil women within a context of post-disaster recovery. Our approach has been both reflexive and reciprocal (Armijos Burneo and Ramírez Loaiza 2021; Armijos and Ramírez Loaiza 2024) in seeking to understand how the women's experiences of the world manifest and are constituted through different scalar relationships located across space and place as well as across time.

These diverse methods have the objective of generating 'new ways of interrogating and understanding the social' (Bagnoli 2009: 548) by facilitating a more in-depth comprehension of the realities of communities. Central to our approach is the use of conversations rather than interviews. This is because conversations are a 'non-structured method of gathering knowledge . . . a combination of reflection, story, and dialogue' (Kovach 2021: 56). Our visits to Chuk Muk occurred between February 2023 and September 2024. We used a diverse range of creative and participatory methods such as social cartographies, drawing and printing with plants, conversations, collective workshops, food preparation, walks, visits to home gardens and building a seed library (Colectivo Miradas Críticas del Territorio desde el Feminismo 2017; Jiménez 2019; Armijos and Ramírez Loaiza 2024). This chapter is mostly based on a series of workshops and our 'conversations from the plants' in June 2024, a methodology created for this project and built around plants.

For these conversations, women were invited to share their journeys along with a plant of their choice and explore stories and memories of plants, places and people. As the women were describing the material part of the plant, one of us was drawing the plant and asking questions about it. We then discussed its importance, relevance in their journeys, care practices and envisioning the future, linking them to the roots, stem, leaves, flowers and fruits. Our participatory methodologies also included productive activities. As we were discussing seeds, plants and food, it emerged that not all the women had access to non-corporate seeds,[2] and thus, the research project responded to this need by providing ten diverse types of seeds[3] to the *Ixquii' Ajtz'utjilaa'* group. They have subsequently been experimenting with different mixes of soil, water quantities, light exposure and the like as part of the workshops.

Finally, as part of our conversations, there was no direct or explicit question about the landslide or the war; however, they were often mentioned in our talks. This decision, informed by a feminist decolonial approach (Armijos Burneo and Ramírez Loaiza 2021; Curiel 2021; Zaragocin and Caretta 2021; Armijos and Ramírez Loaiza 2024), was taken to focus the conversations on what has happened after the disaster and to avoid triggering strong emotions through our questioning. Because the people they lost to the mud and the armed conflict are an important part of their memories and bodies, women would speak about them during conversations about their daily work or other topics we discussed. For example, several times when we met while it was raining, the women would comment on how the weather and the day were very similar to the days before the landslide when rainfall accumulated on the slopes of the volcano, subsequently triggering the flow. They would also mention that they feel safe in the places where they live now. Even if they struggle in their new homes and landscapes, they do not want to go back.

Spaces for Recovery: (Re)making Home Gardens

In this section, we argue that home gardens are spaces where women reconnect with land, memories and traditions around cultivation and food, which enables them to re-territorialize in their (new) homes and landscapes. Chuk Muk, the site where these women were resettled, is approximately 6 kilometres from Panabaj, where the women lived before (Figure 6.1). It was built on a mosaic of lands, over rediscovered archaeological artefacts, and in an area deemed less landslide risky. Despite the seemingly short distance between their old homes and the new settlement, the change was massive and disrupted people's livelihoods and lives in important ways (Cabrera Pacheco et al. 2024). Norris et al. (2019) show that recovery is undermined when families must adapt to smaller plots of land, alongside new neighbours and limited income opportunities in resettlement towns.

The average plot each household received in Chuk Muk has a surface of 150 square metres (Aguirre Cantero 2011). The area included 40 square metres for a vegetable garden. However, the space for living was very limited – only two bedrooms, when Tz'utujil families are traditionally more numerous. Therefore, all the available space, including garden space, is generally used for extra rooms or for relocating other spaces, such as the kitchen. As Esperanza,[4] one of the women in Chuk Muk with whom we worked, said, 'what we do not have here is space to plant'. However, as she reflects, '[even if] the space they gave us here is small, we have to make the most of it, to accommodate to the space to see what can be done. We have to see each space as an opportunity for planting something'.

Women in Chuk Muk have had to create, often out of very small areas, their own spaces for cultivating food. Some, like Esperanza and Carmen, are using communal green spaces next to their plots that were not being used to create their gardens; others, like Sonia and Mujer de Sábila, are extending the planting area by having pots hanging from the windows or using crates or pots on their terraces. Lucia, for example, asked a neighbour for a little space for her lime tree as hers is already used for other purposes. In these small home gardens, women are recreating spaces where they can re-territorialize and maintain a relationship with plants and land. These small forms of cultivation are often disregarded in terms of knowledge and production because of their scale and the fact that their purpose is 'domestic' (Howard 2014). Nonetheless, within home gardens, however small, women keep a vast number of crop varieties, usually native, grown for food and medicinal purposes as well as for ornament (Cabrera Pacheco 2022, 2024). In these unique and intimate spaces, knowledge grows as the plants grow, the women realizing positive spatio-temporal relationships through formative experiences driven by their own very personal motivations, ultimately composing their own re-territorialization in place.

As Shillington (2008) argues, in post-disaster contexts, home gardens facilitate women reconnecting with the land. Home gardens are a space where women can shape the nutrition and health of their families, while also shaping these spaces through a careful layering and positioning of the different trees, crops, herbs and flowers. This became evident in one of our first visits to their home gardens (February 2023) when we asked the women what plants they were growing in their gardens and why. Among others, María Ixaq Sqaaj, on the corridors beside her house, had herbs (*chipilín* and *hierba blanca*), coriander, celery, radish, carrots, coffee, bananas, lime, corn and nance fruit. Rosaura, in her backyard, had *güisquil*, cabbage, beetroot, radish, chillies, lime, papaya, rue, pomegranate and lemongrass. Carmen, in the communal green space next to her house, had coriander, cabbage, onion, celery, banana, chillies, avocado, *míspero*, *güisquil*, epazote, lime, rue and *chipilín*. She also has a larger space for milpa (corn, squash and beans). Sonia, in her small backyard and hanging from pots, had lettuce, chard, radish, beetroot, cabbage, spinach, onion, mango, rosemary, *jocote*, lime, dragon fruit, banana, coffee, *maxán*, *güisquil* and *pacaya*. These plants are important for sustaining everyday activities of care and social reproduction such as cooking and medicinal purposes. They are at the heart of

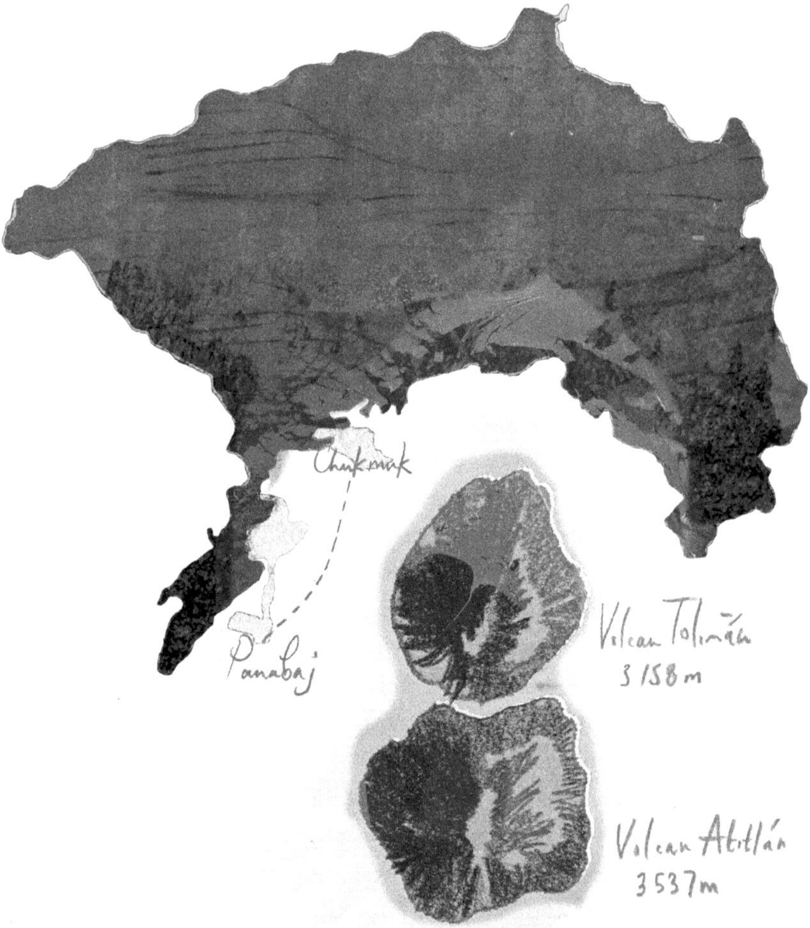

Figure 6.1 Map of Panabaj and Chuk Muk, showing the Lake Atitlán and the volcanoes. *Source: Lisa Mackenzie.*

Tz'utujil cultural practices and traditions and are an important element of how people live in and with their territories. For example, Pilar shared with us why she grows plants in her home garden: 'It is a joy, because you see the fruit of your work. Watching it grow is a motivation to continue cultivating more and more. It is also another reason to eat what you cultivate, as they are healthier dishes'. Rosaura said: 'It makes me feel good, I like what I see in my backyard.

It is a joy to see my plants grow. I feel proud of myself because I see the result of my efforts'.

Having a home garden (Figure 6.2), even if small, also means being in connection with the teachings of mothers and grandmothers and remembering those who are no longer around. Memory can be constructed as a collective and reflexive knowledge, valuable for traditions, such as cultivating and cooking, as it is a knowledge that is transmitted orally through generations. When memories intertwine with traditions, knowledge and practice, they strengthen the agri-food strategies of Maya women (Cabrera Pacheco 2022, 2024) and allow them to navigate these new spaces through the plants. For example, after receiving her seeds and taking part in the workshops, Lucia expanded her small home garden in a covered and secured area in her backyard. She has harvested herbs a few times since and reflected:

> Working with the land provides us with food. Having my own garden has been a major help. It helps for my family to eat healthy food. . . . We are already eating the *chipilín*, and the *bledo* as well. My children are happy when they get home and cut the herb. We mix it with eggs. Not only are we learning to take care of the plants, but we are recovering dishes from the past.

Lucia shared with us her recipe for *hierba mora* or nightshade (*Ich'aj pi'x cuyaa*), which she learnt from her mother and grandmother. She learnt to prepare these dishes since an early age so she could take care of her family in the future. The main ingredient is *hierba mora*, but it can also be made with *chipilín*, *hierba bledo* and chard, whatever they have access to in their gardens or in the market. Nowadays, she prepares this food every week, as she has the herb in her home garden and accompanies it with some chillies from her harvest. When her family has some resources, they get some fish to eat with it. A traditional dish for the Tz'utujil people is *P'atin q'anqaal*. Elena, who is already a grandmother, learnt this recipe from her own grandmother. One of the main ingredients for wrapping up the *patin* is the maxán leaves, which she had in her garden in Panabaj. Now she does not have enough space for them, so she has to buy the bundles at the market. This shows that while having a small garden is important for several reasons, it is still not enough.

Within recovery processes, some authors (see Taiban et al. 2020) have emphasized the importance of local culture for building, and rebuilding, resilience among resettled communities. Rituals, social organizations and local crops are all part of this culture. As we have shown, traditional crops are transformed into traditional dishes by these women. As María Ixaq Sqaaj shared with us, 'My mom taught me how to cook with herbs. . . . She taught me that I should not worry about other recipes but always have herbs nearby. [Now] I have herbs [*chipilín* and *hierba blanca*] here in my home garden.'

Figure 6.2 A growing home garden in Chuk Muk.
Source: Photo by Ana Julia Cabrera Pacheco.

In Chuk Muk, women's intentions for their home gardens are overturning the lack of opportunities in their settlement, particularly for women, to provide healthier food options and readily available medicinal remedies for their families and neighbours and to depend less on the market with its fluctuating prices in food staples. By wanting to generate their own seeds and, with the seed library,[5] conserve and share them, instead of buying seeds every time, the group of women is generating their own resources. With this type of action, they are contributing to the continuation of their Maya way of life while resisting the impacts of the disaster. The women devote their time, energies and agency to practices of initiating slower change and emergence 'with' the ground, in a process of re-territorialization, by co-creating small growing spaces in a respectful encounter with the natural world as they rebuild their connection to land and plants. As with the villagers from the study of Taiban et al. (2020: 12), '[communities] did not derive the energy of spiritual warmth and cultural recovery from safe and durable houses provided by the government. Instead, they healed their wounded hearts by growing traditional crops and adapting through manual labour'. Home gardens in Chuk Muk are allowing for (re)connecting past and present – by means of plants and food, as well as memories and legacies, and by looking towards the future.

Bodies under Recovery: The Plants of Tz'utujil Women

In this section, we show that the intergenerational knowledge of plants that women have is reproduced in their new home gardens through a process of cultivation. Further, we engage with medicinal plants as a process of re-territorialization through women's bodies, particularly after being resettled. It is this displacement that Robin Kimmerer (2020: 16), from the Potawatomi Nation in North America, refers to when she says: 'It separated us from our traditional knowledge and lifeways, the bones of our ancestors, our sustaining plants – but even this did not extinguish identity.' She also acknowledges that 'our people carried [their plants] wherever they migrated' (Kimmerer 2020: 14). In Chuk Muk, women have rebuilt their home gardens with plants that are important to them – they used them in Panabaj and carried, if not the plant, at least the knowledge, traditions and the cultural meanings with them to Chuk Muk. For this group of Tz'utujil women, plants are both 'medicine and a relative, [plants'] value is both material and spiritual' (Kimmerer 2020: 5). For example, Lucia referred to her lime tree as 'a father, a mother, a house partner. It is like a friend, part of the family, because it provides with what I need, it is a help in our home'. When deciding on their important plant as part of our 'conversations from the plants', most of the women chose plants that can be used in traditional medicine, besides being used for eating or drinking. They have first-hand knowledge of illnesses and diseases. Certain plants are important for these women because they have contributed to healing a family member, particularly their own children.

María Ixaq Sqaaj recounts, 'I brought my epazote plant (*sqaaj* in Tz'utujil) from my home [in Panabaj].' She has been outside Panabaj for seventeen years now. She thinks of the epazote as her '*compañera*' (partner) for the future, as it is good for many medicinal remedies. When her son was ill, the elder women in her community advised her to treat him with epazote. She also uses it for stomach aches, mixed with two guava leaves and gramicidin (a type of antibiotic). The leaves are also good for cuts, and if you mix them with hibiscus leaves it is good for body pain. She adds further: 'It is good for the rain or the cold, it warms your stomach.' Elena learnt about lemongrass (*l'muens jyu* in Tz'utujil) from her mother and grandmothers when she was a child. She was separated from her lemongrass when they were in the emergency shelters outside Panabaj for three years. She lost the connection with the plant: 'We missed each other,' she said. She could get it from the market in Santiago Atitlán, but she felt that it was not the same. When she and her family first arrived in Chuk Muk, they had to 'recover' their lemongrass, because 'it is useful, and it is knowledge from our elders'. Her husband collected some lemongrass roots by the lake-shore and brought them to grow in Chuk Muk in a space near the street. She says she is confident in her knowledge when she uses this plant because she has practised and has inherited such knowledge. She stresses that people need dedication and commitment when making a medicinal remedy from this or other plants. Lemongrass is beneficial for fever, headaches, stomach burns, body pain and tonsil infections, mixed with lime, but also for *mal de ojo* (evil eye).

> Sometimes, when children cry or are feverish what you can do is to boil this plant and bathe the child with this water. . . . I only used lemongrass, but my daughter-in-law, what she does with my grandchildren is to mix it with a small twig of cypress in addition to the lemongrass. With that she bathes my grandchildren.

Similarly, Lucia knows that using leaves from the lime tree mixed with guava leaves is beneficial for people with diabetes. Esperanza recommends baths with leaves of avocado, coffee, lime and orange for colds. She says it is good for both kids and women. Mujer de Sábila[6] told us that when her daughter had an accident and burned her legs, she used aloe vera to help her heal. She knew that aloe vera was good for burns from her mother, and she always kept a plant in her house in Panabaj. Now, in Chuk Muk, she keeps many aloes around her house in pots.

While some of these plants serve as remedies for physical illness, others are also useful for mental health and spiritual rituals (Kimmerer 2020). For example, in one conversation, Lucia said that being in Chuk Muk brings back memories of Panabaj and that she feels trapped because now they can no longer access the lake or the volcano slopes in the same way, which sometimes causes her sadness and worry (*b'iis* and *chub'ik* in Tz'utujil). In another conversation, Elena recommended lemongrass for similar feelings:

> [W]hen you feel a heaviness in your chest; when you feel like you cannot breathe with worry, you can make a lemongrass tea with [sodium bicarbonate and acetylsalicylic acid]. With that, you can feel how your chest is relieved, and you can rest. It feels as if it cleanses your heart.

Lucia also shared with us that working in her garden and sharing time and knowledge with the rest of the *Ixquii' Ajtz'utjilaa'* group, who participated in the project, throughout this research process has also helped her feel better.

When women can find the plants that they need in their small home gardens, these spaces become a place for the continuity of situated memories, traditions and knowledges (Gilbert and Williams 2020; Cabrera Pacheco 2022, 2024; Vasudevan et al. 2023). Thus, healing is connected to land through the bodies of Maya women who use herbs and leaves to heal their families. Elena reflected,

> That is why I cannot forget all of this knowledge, because they were words of my grandmothers and my mother. [. . .] If we keep practicing these knowledges we will know from where we came and where we will go. [. . .] Throughout the years I learnt that Mother Earth provides not only with food, but with remedies for all diseases. I learnt that the wisdom of our elders is a richness that we have and that cannot be forgotten because it is for our own development. When I use the plants, I feel like my grandparents and my parents are accompanying me.

These women know that they must adapt, adjust and survive so that they do not lose their roots.

Final Thoughts

The research that we have presented in this chapter set out to understand women's experiences of (re)inhabiting a territory after a disaster. Our research has been guided by both post-disaster and feminist geographies. Post-disaster geography theory guided us to attend to the ways in which we might understand the agency of alternative capacities that arose in the women we worked with in the aftermath and in the ongoing nature of a destructive event. Feminist geographical approaches inspired us to look at the reconstitution of territory through small-scale, intimate practices of women becoming *with* the territory. In unison, the theories combined to challenge the geographical impulse to view post-disaster localities as static, obsolete and exhausted and instead raised questions about their promise as places of exchange and possibility. As post-disaster geographies and feminist geographies evolve and grow through new responses and feedback, it is within these overlapping and shifting boundaries that we have worked together and with the *Ixquii' Ajtz'utjilaa'*, our endeavours in research conjoining with theirs in life.

We have shown the way in which this group of women articulate their present and their Tz'utujil culture by reproducing and reappropriating the plant knowledge from their mothers and grandmothers. These are present in their food recipes and remedies, as well as in the ways those knowledges come about – circulating memories, experiences, feelings and emotions (Ulloa 2021). Through the process of plant growing, the *Ixquii' Ajtz'utjilaa'* are going further by also reclaiming their land. As Sonia said at the beginning of this chapter, to plant directly into the soil means something more to them. Amid everyday struggles, these women have prioritized caring for their families – with healthy food and natural remedies – and caring for their lands. By learning, experimenting and sharing knowledge on soil, seeds, water and seasons they are building appropriate links between their home gardens and their territories. Through the work in these spaces, they have planted the seeds of their own recovery process. They are doing so by reconnecting past, present and future through the generation of cumulative knowledge circles and sharing practices.

During this process, we have, as researchers, also journeyed along with the *Ixquii' Ajtz'utjilaa'*. We have learnt, unlearnt and relearnt (Walsh 2023) not only about the research process but also about the importance of plants in these women's recovery. How they are not objects, but subjects, or relatives, as Kimmerer (2020) has reminded us. Their life journeys through their territories are also the plants' journeys. These plants have helped the women build important connections between their past and present lives. Through their home gardens and their everyday embodied practices (Truelove 2011; Ulloa 2021), they are creating a continuity of their Tz'utujil ways of life that endures.

Acknowledgements

This research was supported by UKRI/GCRF under Grant NE/T010517/1 for the project 'Ixchel: Building understanding of the physical, cultural and socio-economic drivers of risk for strengthening resilience in the Guatemalan cordillera', of which all authors are part. It benefited from additional funding led by Ana Cabrera Pacheco from a UoE RIGLE Small Grant (2023) and the Scottish Funding Council (SFC) International Science Partnerships Fund Institutional Support under Grant 2023-24: SFC/AN/21/2023. We are deeply thankful to the women from the group *Ixquii' Ajtz'utjilaa'* for their wisdom, time and delicious food. We would like to thank Carlos López Nimacachi and Carlos Alfredo Puac for their support during the first moments of this research and Eliza Calder, Alejandra Colom and Lorenza Díaz for their invaluable support in obtaining the additional funding.

Notes

1 The Guatemalan armed conflict (1960–96) has been recognized as genocidal. In Santiago Atitlán, it reached its strongest point in December 1990, with the massacre of fourteen people. Following this, the Tz'utujil community demanded and succeeded in the expulsion of the army from their territory (Comisión para el Esclarecimiento Histórico 1999).

2 For a wider Maya regional context of the appropriation of 'Indigenous grains and territories', see Llanes-Ortiz (2015: 10).

3 Names in Tz'utujil with translations to English (and Spanish in some cases): *Pixnek'* (nightshade – *hierba mora*), *keq ruxee'* (radish), *sirsik ichaaj* (cabbage), *tees* (amaranth), *skaj* (epazote plant), *klaantr* (coriander), *muuch'* (chipilin plant), *pepin* (cucumber), *culix* (white herb – *hierba blanca*), *awen* (corn).

4 The women chose their own pseudonyms for this research project. Some chose plant names, in both Spanish and the Mayan Tz'utujil language. We chose pseudonyms for those who were not available to decide. Quotes originate from our conversations and workshops in 2023–4.

5 We developed an MSc research project, led by Marie Vendrome, under Ana's supervision, on seed libraries. This project was agreed in advance with the group of women and involved their participation in a series of workshops and conversations around the conservation of seeds. As part of the process, the women, Marie and Diego designed and built a seed library.

6 The name she chose translates to 'Aloe Woman'.

References

Adams-Hutcheson, G. and P. Smith (2020), 'Skin, Sweat and Materiality: Feminist Geographies of Emotion and Affect', in A. Datta, P. Hopkins, L. Johnston, E. Olson and J. M. Silva (eds), *Routledge Handbook of Gender and Feminist Geographies*, 80–91, Routledge.

Aguirre Cantero, E. (2011), 'La primera ciudad Tz´utujil del siglo XXI', in *Reasentamiento preventivo de poblaciones en riesgo de desastre Experiencias de América Latina*, 159–92, Banco Mundial & GFDRR.

Armijos, M. T. and V. Ramírez Loaiza (2024), 'Creative Research with Indigenous Women: Challenging Marginalisation Through Collective Spaces and Livelihoods Practices', *Disaster Prevention and Management: An International Journal* 34 (1): 117–32. https://doi.org/10.1108/DPM-01-2024-0037.

Armijos Burneo, T. and V. Ramírez Loaiza (2021), 'Encuentros y Desencuentros con la Fórmula del Riesgo y los Desastres: Tiempos para el Cuidado y la Reciprocidad', *Revista de Estudios Latinoamericanos Sobre Reducción Del Riesgo de Desastres REDER* 5 (2): 19–33.

Bagnoli, A. (2009), 'Beyond the Standard Interview: The Use of Graphic Elicitation and Arts-Based Methods', *Qualitative Research* 9 (5): 547–70. https://doi.org/10.1177/1468794109343625.

Bonilla, Y. (2020), 'The Coloniality of Disaster: Race, Empire, and the Temporal Logics of Emergency in Puerto Rico, USA', *Political Geography* 78:102181. https://doi.org/10.1016/j.polgeo.2020.102181.

Bradshaw, S. and M. Fordham (2015), 'Double Disaster: Disaster Through a Gender Lens', in *Hazards, Risks, and Disasters in Society*, 233–51, Elsevier. https://doi.org/10.1016/B978-0-12-396451-9.00014-7.

Cabrera Pacheco, A. (2022), 'La (re)producción de las tradiciones agroalimentarias en los solares mayas de la Península de Yucatán, México', *Estudios Sociales. Revista de Alimentación Contemporánea y Desarrollo Regional* 32 (60): 1–31. https://doi.org/10.25836/es.v32i60.1263.

Cabrera Pacheco, A. (2024), 'Traditions and Territories of Maya Women in the Peninsula of Yucatán', *Gender, Place & Culture* 32 (5): 775–95. https://doi.org/10.1080/0966369X.2024.2443210.

Cabrera Pacheco, A., J. Cupples, C. Gleghorn, C. A. Puac and R. Ribeiro (2024), *Desastre, reconstrucción, liderazgo: Sistematización de las experiencias tras el deslave de Panabaj de 2005*, University of Edinburgh. https://doi.org/10.2218/ED.9781836450658.

Camacho Benavides, C. (2020), 'Conformación, transformación y resistencias de un sistema biocultural alimentario. El caso de Chiltoyac, Xalapa, Veracruz', PhD thesis, UAM.

Colectivo Miradas Críticas del Territorio desde el Feminismo (2017), *Mapeando el cuerpo-territorio. Guía metodológica para mujeres que defienden sus territorios*, Colectivo Miradas Críticas del Territorio desde el Feminismo.

Comisión para el Esclarecimiento Histórico (CEH) (1999), *Guatemala, memoria del silencio*, Oficina de Servicios para Proyectos de las Naciones Unidas (UNOPS).

Curiel, O. (2021), 'Constructing Feminist Methodologies from the Perspective of Decolonial Feminism', in Y. Espinosa-Miñoso, M. Lugones and N. Maldonado-Torres (eds), *Decolonial Feminism in Abya Yala: Caribbean, Meso, and South American Contributions and Challenges*, 43–59, Rowman & Littlefield.

Fernández, A., J. Waldmüller and C. Vega (2019), 'Comunidad, vulnerabilidad y reproducción en condiciones de desastre. Abordajes desde América Latina y el Caribe', *Íconos – Revista de Ciencias Sociales* 66:7–29. https://doi.org/10.17141/iconos.66.2020.4156.

Gilbert, J. L. and R. A. Williams (2020), 'Pathways to Reparations: Land and Healing Through Food Justice', *Human Geography* 13 (3): 228–41. https://doi.org/10.1177/1942778620951936.

González-Ramírez, A. M., F. Güiza, M. S. Reyes-Quintero, M. E. Méndez-López and P. Torres-Lima (2022), 'Vulnerabilidad Socioambiental desde los Feminismos Comunitarios: El Desastre por el Sismo del 19S en Tetela del Volcán, Morelos, México', *Revista de Estudios Latinoamericanos Sobre Reducción Del Riesgo de Desastres REDER* 6 (1): 35. https://doi.org/10.55467/reder.v6i1.83.

Haraway, D. J. (2016), *Staying with the Trouble: Making Kin in the Chthulucene*, Duke University Press.

Howard, P. (2014), 'Ethnobotanique et Cuisine: L'Importance de la Sphère Domestique Pour l'Agrobiodiversité', in H. Guétat-Bernard and M. Saussey (eds), *Genre et savoirs: Pratiques et Innovations Rurales au Sud*, IRD Éditions. https://doi.org/10.4000/books.irdeditions.9125.

Jiménez, D. (2019), *Geo-grafías comunitarias. Mapeo Comunitario y Cartografías Sociales: Procesos creativos, pedagógicos, de intervención y acompañamiento comunitario para la gestión social de los territorios*, Camidabit-Los Paseantes.

Kimmerer, R. W. (2020), *Braiding Sweetgrass: Indigenous Wisdom, Scientific Knowledge and the Teachings of Plants*, Penguin.

Kovach, M. (2021), *Indigenous Methodologies: Characteristics, Conversations, and Contexts*, 2nd edn, University of Toronto Press.

Llanes-Ortiz, G. (2015), 'Seeds of Maya Development: The "Fiestas y Ferias de Semillas" Movement in Yucatan', *Alternautas* 2 (2): 10–20. https://doi.org/10.31273/alternautas.v2i2.1010.

Massey, D. (2013), *Space, Place and Gender*, John Wiley & Sons.

Naylor, L., M. Daigle, S. Zaragocin, M. Ramírez and M. Gilmartin (2018), 'Interventions: Bringing the Decolonial to Political Geography', *Political Geography* 66:199–209. https://doi.org/10.1016/j.polgeo.2017.11.002.

Norris, L., K. McNamara, N. Fernando, R. Westoby, D. Abeywardhana, P. Vijekumara and L. Prasanna (2019), 'Resettlement Stories of Loss and Barriers to Recovery Following the 2016 Aranayaka Landslide, Sri Lanka', in *Equitable Resilience*, National Building Research Organisation 10th Annual Research Symposium, Sri Lanka.

Shillington, L. (2008), 'Being(s) in Relation at Home: Socio-natures of Patio "Gardens" in Managua, Nicaragua', *Social & Cultural Geography* 9 (7): 755–76. https://doi.org/10.1080/14649360802382560.

Smith, L. T. (2022), *Decolonizing Methodologies*, 3rd edn, Bloomsbury.

Sou, G. (2022), 'Reframing Resilience as Resistance: Situating Disaster Recovery Within Colonialism', *Geographical Journal* 188 (1): 14–27. https://doi.org/10.1111/geoj.12413.

Taiban, S., H.-N. Lin and C.-C. Ko (2020), 'Disaster, Relocation, and Resilience: Recovery and Adaptation of Karamemedesane in Lily Tribal Community After Typhoon Morakot, Taiwan', *Environmental Hazards* 19 (2): 209–22. https://doi.org/10.1080/17477891.2019.1708234.

Truelove, Y. (2011), '(Re-)conceptualizing Water Inequality in Delhi, India Through a Feminist Political Ecology Framework', *Geoforum* 42 (2): 143–52. https://doi.org/10.1016/j.geoforum.2011.01.004.

Tzul Tzul, G. (2015), 'Mujeres indígenas: Historias de la reproducción de la vida en Guatemala. Una reflexión a partir de la visita de Silvia Federici', *Bajo El Volcán. Revista Del Posgrado de Sociología* 1 (22): 91–9.

Ulloa, A. (2021), 'Repolitizar la vida, defender los cuerpos-territorios y colectivizar las acciones desde los feminismos indígenas', *Ecología Política* 61:38–48.

Vasudevan, P., M. M. Ramírez, Y. G. Mendoza and M. Daigle (2023), 'Storytelling Earth and Body', *Annals of the American Association of Geographers* 113 (7): 1728–44. https://doi.org/10.1080/24694452.2022.2139658.

Vergara-Saavedra, P. (2024), 'States Facing Socio-Natural Disasters: Progress and Issues in Latin America and the Caribbean', in A. Farazmand (ed.), *Global Encyclopedia of Public Administration, Public Policy, and Governance*, 1–8, Springer.

Walsh, C. (2023), *Rising up, Living on: Re-existences, Sowings, and Decolonial Cracks*, Duke University Press.

Watts, M. and K. Brannum (2023), 'Bomberos, maestros y psicólogos: Guatemalan civil society response to the Volcano of Fire disaster', *Journal of Human Rights* 22 (2): 252–67. https://doi.org/10.1080/14754835.2022.2115288.

Zaragocin, S. and M. Caretta (2021), 'Cuerpo-Territorio: A Decolonial Feminist Geographical Method for the Study of Embodiment', *Annals of the American Association of Geographers* 111 (5): 1503–18. https://doi.org/10.1080/24694452.2020.1812370.

7

The Affective Politics of Magma in Andean Worlds

Navigating More-Than-Human Kinship with Volcanoes in Disaster Research

Francisca Vergara-Pinto

The Sierra Nevada is just over there. It is always here in the village looking down on us. . . . The Lonquimay volcano is my father and the Sierra Nevada is my mother. For that reason, I could never leave this land.

– Resident and ski patrol, Malalcahuello, Chile

This chapter draws on geographical theories of affect to advance conceptualizations of volcanic kinship and the affective politics of magma, in which people do not merely coexist with volcanoes but engage with them as familial formations, ancestral beings and more-than-human agents. This kinship reflects a relationship with volcanoes in which they serve as a foundation for human life while representing a risk to such life through the disasters they may trigger. Active volcanism, whereby magma rises to the surface to form volcanoes, is one of the most ubiquitous geological forces in the material creation of habitable worlds. In fact, more than 1 billion people live within 100 kilometres of a volcano (Freire et al. 2019), a proximity that is closely connected to cultural traditions in many locations around the world (Calvache and Sánchez 2022). The presence of volcanoes in art, music, literature and the histories of various cultures suggests that nearly every volcano worldwide is linked to humanity's cultural diversity (Cashman and Cronin 2008; Acuña et al. 2023).

Thus, volcanoes have been continuously associated with life, from forming the primordial atmosphere to shaping today's landscapes, which are used for everything from tourism to the fertile soils that provide resources, food and well-being to sustain communities (Sheets and Grayson 1979). These widely used benefits have been essential for the reoccupation of volcanic lands after eruptions, despite the risk of destruction these spaces represent. The idea and evidence put forward by biologists and geochemists – that life could not have emerged without volcanism (Wächtershäuser 2006) – are compelling. This idea is consistent with an intriguing issue for anthropology and human geography (e.g. Cannizzaro 2018; Faas 2023; Vergara-Pinto and Marín 2023): there are people who feel that *they could not live without volcanoes*. In addition to the benefits and risks, there appear to be other factors behind the persistence of human life on volcanic lands, more closely related to affective dimensions and cultural elaborations of what a volcano is.

Reoccupying volcanic lands comes with significant benefits but also with challenges, concerns and contradictions. Choosing to live near a volcano or to return to it after an eruption raises important questions about belonging, displacement and humanity's capacity to withstand overwhelming more-than-human forces. These decisions made by residents ultimately touch on matters of life and death, whether through experiencing an eruption or expecting it as something in the making, unavoidable. Due to the intensity of their behaviour, volcanic eruptions are among the most powerful, destructive or even deadly natural hazards. The 79 AD eruption of Vesuvius is a remarkable example, which destroyed the city of Pompeii in Italy and has since been reoccupied (Sigurdsson et al. 1982). In many other locations, humans have also been confronted with loss, death, hope and the need to repair their worlds with the volcano after the destruction (Ramírez et al. 2017). Because of this, volcanoes have the potential to affect and elicit ambivalent emotions in humans. In this sense, magma came first, then came life, and today human life and volcanism come together. As a result of the rhythm of life they create alongside volcanoes, communities sustain an affective connection with them in their different possibilities, *moods* and circumstances. This connection is manifested and embodied through feelings, practices and deep senses of belonging that, far from being anecdotal, are formed throughout a person's life, across generations and in tune with fragments of the geological life of a volcano.

Belonging to a volcano as an affective bond takes manifold forms. In particular, because of their association with the creation of life and the reproduction of cultural identities, many volcanoes around the world are known by people living close to them as a mother (*Mama* Tungurahua, *Mamma* Etna), grandfather (*Taita* Chimborazo; Lyons 1999; Della Dora 2016; Faas 2023) or daughter (*North* and *South Daughter* of the Rabaul caldera; Epstein 1992). In other cases, there are *children* of a volcano (children of Chã das

Calderas; Weeks and Popinsky 2016) or *people* of a volcano (Tenggerese and Llaimache peoples; Setiawan et al. 2022; Dowling 2023), among other familial denominations. Kinship, understood as a way of 'making kin' and establishing 'correspondence' with other humans and more-than-humans (Viveiros de Castro 2009; Sahlins 2011; Haraway 2016; Ingold 2017), is therefore present in the complex relationships people have with volcanism. Drawing on research into narratives of local communities living in the Andes Cordillera in South America, this chapter asks: What characterizes this volcanic kinship?

With this chapter, I intend to advance the integration of volcanoes and geographies of affect to understand forms of more-than-human kinship in volatile contexts and their role in underpinning and organizing disaster risk. On this basis, I reflect on the influence that the eruption of a volcano, considered a family relative on a daily basis, can have on people's awareness and understanding of hazards during situations of volcanic crisis. Ultimately, this question urges human geography to provide new insights into how magma becomes a more-than-human entity and to explore how kinship and affect shape the ways people share their worlds with volcanoes.

More-Than-Human Kinship: Insights from the Geographies of Affect

The concept of kinship has traditionally been an anthropological concern focused on social structures, familial relationships and social organization (Lévi-Strauss 1965; Ingold 2017). Recent scholarship connecting anthropology, human geography and the environmental humanities has expanded the idea of kinship to include relationships between humans and non-human beings such as animals, landscapes and geological features. It has thus shifted from being a concept anchored to determinations by genealogies and consanguinity to more performative processes (Sahlins 2011). This notion appears early in Aristotle's philosophy, for whom the 'discussion of kinship encompasses relations of procreation in larger meanings of *mutual belonging* that could just as well accommodate the various performative modes of relatedness' (Sahlins 2011: 10). Using the example of the bond between parents and children or siblings, Aristotle says that 'they love them as being themselves. . . . They are, then, the same entity in a way, even though in discrete subjects' (Sahlins 2011). Therefore, kinship emerges among persons who belong to one another, who are 'co-present' in each other, whose lives are 'joined and interdependent'. Drawing on Sahlins's definition, kinship operates as a 'mutuality of being' among beings who are intrinsic to one another's existence.

In this view, kinship is eminently affective, sustained in the co-presence of kin who engender corollaries in the 'transpersonal unities of bodies, feelings and experience' (Sahlins 2011: 11). In this light, the phenomenon of belonging – to a person, a place, a culture or a thing – and its transfer to others through kinship becomes relational. That is, kinship is not to be sought as content or representation in an individual but as a force in the bond that is co-created and embodied between people and things. How, then, should we approach socio-cultural realities in which an animal, a plant, a river or a volcano enters into this fabric of bonds? And how can we understand such a sense of belonging when it simultaneously involves a natural threat? Certainly, ethnographic evidence shows that many 'lands' or 'places' that are often simply understood as contexts are, in fact, held by communities as a non-human 'entity'. As such, kinship relations are not only 'in' that specific place but also 'with' that place. According to recent research on the affective connection with volcanic lands, a volcano becomes a geological feature with agency and influence in social life, as it articulates a set of situated relationships that reinforce belonging with the land and 'across species' (e.g. Cunningham 2018; Desai and Smith 2018; Morin de Lima and Soares-Pinto 2024; Vergara-Pinto et al. 2024). Such a relational understanding paves the way for reflection on how kinship, as a classic anthropological concern, might be brought into dialogue with geographical theories of affect.

Affect refers to forces, intensities and shared embodied responses to changing environments as a mode of sense-making (Massumi 1995; Seigworth and Gregg 2010; Anderson 2014). For example, in extreme events, people may respond with a flow of affect emerging between their bodies, from dread to hope, from fear to excitement (e.g. Anderson 2020). These affects take place as precognitive sensations that arise within an individual, often preceding conscious emotions. They act as ways in which bodies 'make sense of, express themselves within and performatively shape different spatially and temporally bound situations' (O'Grady and Shaw 2023: 518). Affects also communicate with the environment, inhabiting the 'passage between contexts' in a flow of attunement with other human and non-human agents such as geological structures (Cloke and Conradson 2018). This creates a reciprocal interplay of relations characterized by a 'distributive agency' of various vitalities at play (Bennett 2010: 32), emphasizing our ability to affect others while also being affected. In this way, it is understood that 'the human body is what it is because of its unparalleled ability to co-evolve with things' (Thrift 2008: 10), surrounded by vital forces and non-human agents that co-constitute the very fabric of human existence (Bennet 2010).

Geographies of affect facilitate a serious exploration of our immediate lived experience of the world (O'Grady 2018), moving us to conceptualize the world beyond its representations. It argues that placing too much emphasis on representation can diminish the richness of our experiences, feelings and ways of being in the world, through which affective interactions between human and non-human bodies take place. Theories of affect therefore diverge from representational approaches by emphasizing the embodied nature of affects and emotions. The agency of a volcano, in particular, produces affects that reflect the connection between the earth and people (Vergara-Pinto et al. 2024). This phenomenon is deeply connected to the relationship between the body and the imperceptible forces and powers that influence it (Deleuze and Guattari 1980; Deleuze 1998) – in this case, the geological forces and powers of volcanism. At this point, we can glimpse the affective nature of more-than-human kinship, defined as 'becoming a human body through the differential bodily engagement of and/or with other bodies, human as well as nonhuman' (Viveiros de Castro 2009: 243).

In geodynamic contexts, the more-than-human aspect of natural phenomena intersects with hazards and risks. When a disaster occurs, people's relationship with nature and their lands is disrupted by the extreme forces of the earth. However, certain factors can enhance and strengthen an individual's sense of belonging to the affected areas. These factors are neither purely human nor purely non-human; rather, they emerge from the relationships between people, places, objects and entities. In this context, kinship networks have been shown to contribute to intra-community resilience in the aftermath of disasters (Oracion 2015; Cajilig 2022), serving as a mobilizing force to help restore life. In these communities, the focus is on maintaining and passing down relationships with relatives at all costs, as the bonds people share with others can provide strength to cope with, recover from or ideally prevent disasters. However, there have also been instances where political decisions have disrupted these relationships rather than the natural event itself, such as volcanic eruptions. A significant example is the 2008 eruption of Chaitén volcano in Chile, where forced relocation from the town (mostly rural) to other cities led to a myriad psychological, health, economic and social issues. This case is frequently studied as an example of how disaffected disaster management can create a crisis by shattering the bond between people and the volcanic places where they live (e.g. Sandoval et al. 2014).

Viewing relationships with more-than-human entities as socioecological kinship networks can enhance our understanding of how communities relate to a volcano and the areas affected by eruptions. Kinship helps to reveal how volcanoes play a significant role in people's lives and personal histories. Indeed, it helps to identify how volcanoes act as agents that both give and take, offering countless benefits when it is calm and posing threats during unrest and eruptions. This dynamic, wherein volcanoes both sustain and endanger people, shapes not only the landscape but also ways of life. However, this topic has received little attention in existing literature. For example, Madera (2021) captures the affective nature of kinship in the way people embody and express their experiences with the Guagua Pichincha volcano in Ecuador. She writes, 'I was raised by volcanoes, and in my sorrow and worry, I turn for comfort and inspiration to the mountains circling around me. . . . I was born on the flanks of this mountain, as was my daughter' (Madera 2021: 19). We can now consider how belonging and kinship are transferred, creating a continuous process of change in which both humans and more-than-humans influence and shape one another.

Against this background, the rest of this chapter employs the notion of more-than-human kinship and theories of affect to explore people's understanding of volcanic kinship. Examples are drawn from several volcanoes located in Latin America, focusing specifically on two countries across the Andean worlds (De la Cadena 2015), shaped by the Andes Cordillera: Ecuador and Chile.

The Andes Cordillera: Lands of Volcano Peoples

I review two references that exemplify the case of Ecuador, focusing specifically on Tungurahua volcano (Parra 2022; Faas 2023), as well as audiovisual sources concerning Chimborazo volcano (DW Documentary 2018; Doctv Ecuador 2020; Día a Día Teleamazonas 2024). Additionally, I revisit an interview I conducted during my ethnography work at Lonquimay volcano in 2022 (Vergara-Pinto et al. 2024). Through this exploration, I illustrate the residents' kinship with the volcano and the familial connections that exist between Lonquimay and other volcanoes in the region. In doing so, I aim to show how people feel and narrate their sense of belonging to active volcanoes.

Ecuador: Northern Volcanic Zone

Ecuador boasts a diverse and dynamic landscape, deeply influenced by its geology and the ongoing subduction process (Troncoso and Ordóñez 2023). Situated within the North Volcanic Zone, the Andes Cordillera in Ecuador stretches from the Chimborazo at Carchi provinces, offering landscapes uniquely shaped by volcanic activity. To date, twenty-seven potentially active volcanoes have been identified in the country. Eruptions have provided essential resources such as fresh water, thermal and mineral-rich springs and fertile soils that sustain agriculture. The unique natural advantages of volcanic regions have historically drawn communities to settle in their proximity, leading to the occupation of these lands despite the inherent risks of eruptions. This pattern has persisted for centuries, even before Spanish colonization, with communities facing eruptions that resulted in the destruction of towns that were courageously rebuilt time and again (Petit-Breuilh 2004). Two volcanoes – Tungurahua and Chimborazo – exemplify this human phenomenon that has deep roots in affects with more-than-human entities and geological otherness.

Mama Tungurahua: The Volcano Grandmother

Tungurahua ('*throat of fire*' in the Quechua indigenous language) is a stratovolcano standing 5.020 metres above sea level and is located between the Tungurahua and Chimborazo provinces. The recent history of Tungurahua volcano is marked by significant eruptions between 1641 and 1646, 1773 and 1781, 1886 and 1888, 1916 and 1918 and the prolonged activity from 1999 to 2016 (Troncoso 2022). The last event garnered significant interest from researchers, who explored the intricate connections between local communities and the revered figure known as Mama Tungurahua (Figure 7.1). From a sociology perspective, Parra (2022) reports on the influence of women's knowledge and the way Mama Tungurahua's conception of women shapes local risk management, expressed through coping strategies based on care, resistance and adaptation. Meanwhile, Faas (2023) provides an anthropological perspective, delving into the beliefs, traditions and cultural practices that embody the relationship between the people and Mama Tungurahua, reflecting their respect for a more-than-human family figure. Together, these studies illuminate the profound ways in which a sense of belonging to volcanoes may intersect disaster risk management.

Research that has examined the community perspectives of the Tungurahua volcano reveals how people commonly designate it as a 'mother' figure. One

Figure 7.1 'Mama' Tungurahua volcano.
Source: Photo by Liliana Troncoso.

understanding of why 'she' is considered a mother comes from religious beliefs. There is a noted connection between Mama Tungurahua and the Virgin Mary, both revered as maternal figures capable of healing their children through holy water, demonstrating people's anthropomorphic view of the volcano. Given the religious significance and healing capabilities of Mama Tungurahua, the town of Baños, situated on the slopes of Tungurahua, has become a significant religious and tourist attraction, with people coming to bathe in the volcanic waters. Specifically, the vibrant hues of the waters in which both indigenous people and later Spanish settlers bathed led people to believe the waters were the urine of Mama Tungurahua, blessed by the Virgin Mary and filled with healing properties (Parra 2022). This belief persists in the community today, where women tend to refer to Tungurahua not simply as a volcano but as Mama, Grandmother or Mrs Tungurahua. The specific term used varies depending on the area of the volcano from which residents view it. As noted by an interviewee quoted in Parra (2022: 84), 'the community on the western side considers her the grandmother. They speak to her as the grandmother and always refer to her as the eldest', considering Tungurahua the elder figure of the entire territory.

Faas (2023) addresses the kinship relationship with Mama Tungurahua in an explicitly affective sense, going beyond reducing it to an anthropomorphic entity. For him, this kinship reflects a reciprocal relationship between human and volcanic materials, in which people see the volcano as a geological being with intentions and influence in their lives:

> The volcano, Tungurahua, was kin – *mama* – since just about forever, though we'll later see how she aged with her children and grandchildren. But she was combustible, mercurial kin – in society, yes, but no anthropomorphic character. Other hills around Tungurahua were consequential social actors, some of which offered aid in life-or-death situations. The boundaries were not only ambiguous and contingent, they simply meant very little. Humans and nonhumans might be spoken of as nature or culture depending on how and when people spoke of them, not because of enduring and immutable characteristics of *what* they were speaking about.
>
> – Faas (2023: 14)

Mama Tungurahua is regarded as a protective yet unpredictable force, embodying both nurturing and destructive aspects. The communities attribute volcanic activity, such as eruptions, to the moods or responses of the volcano to people's *good* or *bad* behaviour. This perspective fosters a sense of accountability and mutual respect, as residents feel they must live in harmony with their environment to appease or understand the volcano's actions 'in ways consistent with their own values and the good to which they aspire' (Faas 2023: 195). Harmony with the volcanic environment involves making offerings to the earth to express gratitude for sustenance, as well as taking proactive measures for disaster preparedness to support the community during a volcanic crisis. An example of this is the Tungurahua *vigías* group (Troncoso 2022), which consists of villagers organized to live in tune with the rhythms of the volcano that erupted for seventeen years.

Respect for the earth is expressed to maintain this balance, underscoring the relational dimensions of kinship with the volcanic landscape. This sense of belonging between people to their volcanoes is deeply rooted in indigenous and rural traditions, blending cosmological beliefs with practical adaptations to the ever-present risk of living near an active volcano. Kinship with Mama Tungurahua serves as a framework for place-based resilience, whereby, rather than perceiving the volcano solely as a threat, the residents' relational view helps them navigate life in a volatile landscape.

Taita Chimborazo: The Volcano Grandfather

Chimborazo ('snow-capped mountain' in the Quechua language), a potentially active volcano and Ecuador's highest peak, holds profound cultural significance for indigenous Andean communities. In Quechua cosmology, Chimborazo is considered an *apu*, or mountain spirit, and is often viewed as a protective and paternal figure. This kinship positions Chimborazo as both an ancestor and a living entity that participates in the community's well-being (Salomon 1986). Ethnographic research underscores how offerings to Chimborazo are integral to maintaining reciprocal relationships with the mountain. These practices often involve *ch'alla*, a ritual offering of food and drink to honour the *apu* and ensure the agricultural fertility provided by the volcanic land. Chimborazo's snow-capped peaks are also seen as a source of life-giving water, reinforcing its role as a provider and caretaker, as well as a pivotal referent of global change (Sarmiento et al. 2022). Such relationships challenge Western dichotomies of nature and culture, emphasizing instead a kinship in which humans and volcanoes are co-participants in shared worlds (De la Cadena 2015).

To illustrate this affective connection with the volcano, I will present some excerpts from the story of Baltazar Ushca, a resident known as the last ice breaker of Chimborazo, who passed away in 2024. These accounts come from open-access documentaries on YouTube platform that explore Baltazar's life. In this section, I demonstrate how affect between people and active volcanoes is embodied and transcended through kinship throughout a person's lifetime and across future generations. The following quotes reflect three moments in Baltazar's life, each depicted in a different documentary. The first refers to his sense of belonging to Chimborazo:

> I never fail Chimborazo. When it's time to come, we just come. My dad was like a son of Taita Chimborazo – white T-shirt, white hair. Maybe I'm almost like a grandson to taita Chimborazo. For me taita Chimborazo is almost a grandfather. A grandfather. I love him very much. We come to Chimborazo with friends, he accompanies us, we talk.
>
> – Baltazar Ushca (2020), Doctv Ecuador

Meanwhile, the second quote illustrates how the kinship connection with the volcano is reflected within his own family:

> My dad taught me and my brothers to work in the ice (glacier) mines. I am the only ice trader in Chimborazo. My dad, my mum and my brothers used to work

> on the ice. . . . Maybe some of my children or grandchildren would like to work like me on the Chimborazo. When I die, this could disappear. As long as my body allows me, I want to continue working.
>
> – Baltazar Ushca (2024), Día a Día Teleamazonas

The last documentary (Doctv Ecuador 2024) shows that both the ice-breaker trade and the kinship with Chimborazo continue to be sustained through the kinship bond itself. This is because a relative of Baltazar (his son-in-law) took up the practice of collecting ice from the volcano's glacier after Baltazar passed away. The son-in-law had accompanied Baltazar during the last years of his work, after the elder suffered an accident, observing every movement of the chisel in Baltazar's hands, becoming his apprentice while learning to withstand the cold of Chimborazo volcano and, in turn, becoming the volcano's *grandson*.

Chile: Southern Volcanic Zone

Chile lies on the active eastern margin of the Pacific and is part of the South Volcanic Zone, a region shaped by the subduction of the Nazca Plate beneath the South American Plate. Its volcanic arc, the Andes Cordillera, consists of approximately 2,000 volcanoes, of which eighty-seven are considered active according to the Chilean Geological and Mining Survey (SERNAGEOMIN). Most of these volcanic areas are inhabited and reoccupied after eruptions. There have been more than thirty documented eruptions in the past few centuries (Amigo 2021). The Lonquimay volcanic complex, located in the Kütralkura UNESCO Geopark, has a history of eruptions, as well as communities (indigenous, campesinos, settler and migrants) that have reoccupied areas close to or within its hazard zones following eruptions (Moreno et al. 2023; Vergara-Pinto et al. 2024; Romero et al. 2025). Five historical eruptions have been documented at Lonquimay volcano: 1852–3, 1887–90, 1933, 1940 and 1988–90 (Polanco 2010). The 1988 eruption, which started on 25 December and lasted for thirteen months, formed the Navidad (*Christmas*) volcano, a new cone on the slope of Lonquimay. Despite the precursory signals that persisted for months, it was not possible to predict the type of eruption due to the lack of adequate monitoring at the time. Even so, part of the population in Lonquimay evacuated. Residents who remained suffered economic losses and serious health impacts due to the emission of gases, while a wave of 'special interest tourism' emerged, as groups came to observe the new cone formation, including residents, scientists and foreigners (Petit-Breuilh 2016).

Lonquimay: The Volcano Father

Lonquimay (meaning '*nodding*', '*salute to the head*' or '*thick forest*' in the Mapuche indigenous language) is a stratovolcano located in the village of Malalcahuello in the Araucanía region of southern Chile. Standing at approximately 2,865 metres above sea level, Lonquimay is known for the Navidad eruption of 1988, which lasted almost a year. This eruption produced extensive lava flows and ash plumes, significantly affecting local communities and ecosystems. My ethnographic study in the village of Malalcahuello (2022–4), where I also live, revealed compelling narratives about the strong sense of belonging and affects the residents hold for the volcano and its eruptions (see Vergara-Pinto et al. 2024). Notably, two inhabitants expressed a shared perspective on how volcanism is intricately woven into their past, present and future life trajectories. Both express a firm commitment to maintaining their connection with the volcanic land, even after an eruption, and to ensuring this vital bond is passed down to their descendants.

A 94-year-old woman shared how the 1988 eruption of Lonquimay profoundly shaped her life and strengthened her bond with the surrounding rural landscape. She raised her children in Malalcahuello, where they continue to live today. When discussing the 1988 eruption, she expresses a deep sense of gratitude towards the volcano, which she understands as a more-than-human entity intertwined with divine forces, revealing an affect that will undoubtedly influence her outlook on future eruptions:

> The Lonquimay volcano is like us, a living being that suddenly falls asleep for a long time, and has to wake up at some point because it must be hungry, it must want to breathe to get everything out from inside. . . . If lava came out from this side, towards Malalcahuello, it would be dangerous. But Navidad (Christmas cone) spared us from such things. The Lord can break wherever he wants, as he did with Navidad (Christmas cone). God prepared and broke the earth there so that this volcano could form next to Lonquimay.
>
> – Resident interviewed in 2022

In territories like Malalcahuello, which is bordered by several active volcanoes including Lonquimay (Figure 7.2), residents skilfully navigate their lives amidst the uncertain, shifting and unpredictable nature of volcanism – oscillating between stillness, unrest, eruptions and calm. Alongside Lonquimay stands the active Tolhuaca and the Sierra Nevada, considered 'extinct'. There is also the

Figure 7.2 View of Lonquimay (left) and Tolhuaca (right) volcanoes.
Source: Photo by Francisca Vergara.

Navidad cone, understood as the 36-year-old child of Lonquimay. Together, these volcanoes form a geological family that envelops Malalcahuello village. Many inhabitants integrate these geological beings into their own familial structures, transforming the concept of kinship into a network of connections that intertwine humans with the more-than-human world. The final example presents the viewpoint of a local resident, a forty-year-old man born and raised in the village. His account highlights how his affective relationship with Lonquimay volcano shapes his perspective on a future eruption. Ultimately, this deeply influences his decision about whether or not to leave his *volcano father* behind:

> I go to the volcano to look for answers, to face problems by turning to nature as an emotional release and feeling the emotion of the top of the volcano. . . . The Lonquimay volcano is my father and the Sierra Nevada my mother. They surround us like giant guardians. For that reason, I could never leave this land.
>
> – Resident interviewed in 2022

As can be seen, kinship performed by people in volcanic places can be understood as the result of 'the ability to co-evolve with things' (Ingold 2021), in this case with volcanic materials that leave traces on people's senses of belonging and their senses of home and family.

Concluding Remarks: The Affective Tone of Eruptions

Volcanoes constitute an extensive field of interest in disaster research because an eruption can result in destruction, loss and death. Yet how the affective constellation of volcanoes mediates people's sense of belonging in everyday life and their vital decisions about potential eruptions is often overlooked. For instance, much is said about belonging and attachment to place as symbolic or immaterial processes. However, by adopting an affective perspective, it is possible to reframe belonging as a force that draws people's bodies to the volcano, akin to an invisible bloodline that inhabitants seek to inherit. The affective nature of this kinship connection between people and volcanoes can make leaving the erupting land feel like abandoning a family member. For this reason, it can be emotionally arduous for people to evacuate and relocate after volcanic eruptions (e.g. Baxter and Horwell 2015; Bird and Gísladóttir 2018; Dowd 2022; Carvajal 2024; Malas and Tolsá 2024). Ultimately, decisions about whether to relocate, remain near the volcano or return to it after evacuation are shaped by power relations among stakeholders (local communities, decision-makers and authorities, among others), who may differ according to their affective or disaffected connections to volcanoes. These power relations can either restrict or facilitate the 'potential' for kinship connections between people and their volcano that are essential to their cultural identities, livelihoods, ways of being and their future as subjects (e.g. Zemelman 2010; Paredes 2013).

Attention to kinship as an affect-laden relationship enacted by local communities can therefore help to examine what I call the affective politics of magma. Here I consider the relationship between humans and more-than-humans and the catastrophic potential this relationship bears. To provide an understanding of this, I elaborate upon an affective politics of magma, whereby political-affective readings can be made of the everyday and catastrophic events that shape the bond between the magma-turned volcano and the human body–turned son, daughter or grandchild of the volcano. This understanding brings to human geography a view of volcanoes as affective bodies whose interaction with humans generates a material and political force. This force is manifest in how our perception of disaster depends on the relative dynamics of detachment or kinship we hold with volcanoes. The volcano thus becomes an affective counterpoint in the geographies of disaster.

Everyday life can foster a harmonious sense of belonging with volcanoes, while volcanic eruptions may strengthen kinship bonds between people and the volcanoes they live alongside. In both situations, we can observe the emergence of minor gestures or micro day-to-day practices that intensify individual experiences and deepen people's connections with the more-than-human world (Manning 2019). This micro-scale can be seen in residents' practices, such as observing the volcano through the window on a daily basis, educating children in a school that uses the volcano as a regular classroom, or collecting medicinal herbs at the foot of the volcano whenever there is pain, among others. During an eruption, small gestures may take the form of contemplating with astonishment the flow of lava or dedicating a prayer and an offering to calm the volcano. These experiences may precede social changes within affected communities, acting as forces that encourage variation in how people live with the unpredictable nature of volcanism. The vibrant presence of volcanism has enabled people to cultivate a sense of kinship in areas dominated by geological forces. Belonging to the volcano implies memorializing the traces left by eruptions in the form of layers of ash and accounts of coping experiences engraved in space and within the family fabric (Vergara-Pinto and Marín 2023). This chapter presents a series of examples that illustrate these various social, geological, cultural and affective ways in which kinship is performed by people, allowing them to recognize that they not only inhabit volcanic areas but share the world with volcano kin. Addressing this type of more-than-human relationship in risk management, risk communication, knowledge dialogue and disaster preparedness poses a significant challenge that increasingly demands the attention of the geography of affect.

Acknowledgements

I extend my deep gratitude to the residents of Malalcahuello who participated in my research on Lonquimay volcano, as well as to Liliana Troncoso for providing a photograph of Tungurahua volcano. I thank my PhD supervisors, Nathaniel O'Grady and Aurora Fredriksen, for their guidance, encouragement and insight throughout my research. Lastly, I am grateful to volcanoes for creating the worlds we live in.

References

Acuña, V., M. Clunes, S. Riffo and J. Browning (2023), 'Intertwining Volcanoes and Society in Chile Through Arts and Interdisciplinary Connections', *Andean Geology* 51 (2): 271–95.

Amigo, A. (2021), 'Volcano Monitoring and Hazard Assessments in Chile', *Volcanica* 4 (S1): 1–20.

Anderson, B. (2014), *Encountering Affect: Capacities, Apparatuses, Conditions*, Ashgate.

Anderson, B. (2020), 'The Affects of the Disaster', *Political Geography* 78:1–2.

Baxter, P. J. and C. J. Horwell (2015), 'Impacts of Eruptions on Human Health', in H. Sigurdsson (ed.), *The Encyclopedia of Volcanoes*, 1035–47, Academic Press.

Bennett, J. (2010), *Vibrant Matter: A Political Ecology of Things*, Duke University Press.

Bird, D. K. and G. Gísladóttir (2018), 'Responding to Volcanic Eruptions in Iceland: From the Small to the Catastrophic', *Palgrave Communications* 4 (1): 151.

Cajilig, P. G. (2022), 'Kinship as Cultural Citizenship in Post-Disaster Housing Reconstruction: Narratives from Fisherfolk in the Philippines', in A. Lukasiewicz and T. O'Donnell (eds), *Complex Disasters: Compounding, Cascading, and Protracted*, 265–87, Springer Nature.

Calvache, L. M. and J. J. Sánchez (2022), 'Arte, Cultura y Economía Alrededor del Concepto de los Volcanes De Nariño, Colombia', *Revista Geológica de América Central* 66:1–15.

Cannizzaro, S. (2018), 'Geographical and Historical Processes of Human Settlements in the Etna Region: A Person-Place Relation Approach', in L. Antronico and F. Marincioni (eds), *Natural Hazards and Disaster Risk Reduction Policies*, vol. 1, 69–87, IL Sileno Edizioni.

Carvajal, G. (2024), 'Chaitén La Pompeya (2015) by María Paz González: A Catastrophe of Sense', *Latin American Theatre Review* 57 (2): 7–30.

Cashman, K. and S. Cronin (2008), 'Welcoming a Monster to the World: Myths, Oral Tradition, and Modern Societal Response to Volcanic Disasters', *Journal of Volcanology and Geothermal Research* 176 (3): 407–18.

Cloke, P. and D. Conradson (2018), 'Transitional Organisations, Affective Atmospheres and New Forms of Being-in-Common: Post-Disaster Recovery in Christchurch, New Zealand', *Transactions of the Institute of British Geographers* 43 (3): 360–76.

Cunningham, E. J. (2018), 'Nature Interrupted: Affect and Ecology in the Wake of Volcanic Eruption in Japan', *Conservation and Society* 16 (1): 41–51.

De la Cadena, M. (2015), *Earth Beings: Ecologies of Practice Across Andean Worlds*, Duke University Press.

Deleuze, G. (1998), *Essays Critical and Clinical,* Verso.

Deleuze, G. and F. Guattari (1980), *Capitalisme et schizophrénie*, Les.

Della Dora, V. (2016), *Mountain: Nature and Culture*, Reaktion.

Desai, S. and H. Smith (2018), 'Kinship Across Species: Learning to Care for Nonhuman Others', *Feminist Review* 118 (1): 41–60.

Día a Día Teleamazonas (2024), 'Farewell to the Last Ice-Breaker of Chimborazo'. Available at: https://youtu.be/EOEFddzOaA8?si=j712MBtfBpPxqezP (accessed 2 November 2024).

Doctv Ecuador (2020), 'Baltazar Ushca the Last Ice-Breaker (Part 1/5)'. Available at: https://youtu.be/HEvb3X19K40?si=SjN1-zqx63qgyeH4 (accessed 2 November 2024).

Doctv Ecuador (2024), 'Chimborazo's New Ice-Breaker is the Son-in-Law of #BaltazarUshca'. Available at: https://youtu.be/UYKZjBdj3aA?si=hIICn7E2jfOleiMK (accessed 2 November 2024).

Dowd, A. (2022), 'Vanishing Acts: The Crisis of Our Loss of Kinship with the More-Than-Human World', *Journal of Analytical Psychology* 67 (5): 1270–95.

Dowling, F. (2023), 'Rescate Desde la Oralidad El Toltó Pewenche, un Arquetipo Arquitectónico Ancestral, la Herencia de los Ultimos Trashumantes de los Andes del Sur', *Cuadernos del Centro de Estudios de Diseño y Comunicación* (185). https://doi.org/10.18682/cdc.vi185.9502.

DW Documental (2018), 'The Last Ice-Breaker of Ecuador'. Available at: https://youtu.be/mESOZvOgs5k?si=P1f6TElLANdZqhVC (accessed 2 November 2024).

Epstein, A. L. (1992), *In the Midst of Life: Affect and Ideation in the World of the Tolai*, vol. 9, University of California Press.

Faas, A. J. (2023), *In the Shadow of Tungurahua: Disaster Politics in Highland Ecuador*, Rutgers University Press.

Freire, S., A. J. Florczyk, M. Pesaresi and R. Sliuzas (2019), 'An Improved Global Analysis of Population Distribution in Proximity to Active Volcanoes, 1975–2015', *International Journal of Geo-Information* 8 (8): 341.

Haraway, D. J. (2016), 'Staying with the Trouble: Making Kin in the Chthulucene', in *Staying with the Trouble: Making Kin in the Chthulucene*, Duke University Press.

Ingold, T. (2017), 'On Human Correspondence', *Journal of the Royal Anthropological Institute* 23 (1): 9–27.

Ingold, T. (2021), 'Kinship by the Sea', in G. Van Horn, R. Wall Kimmerer and J. Hausdoerffer (eds), *Kinship, Belonging in a World of Relations, Vol. 3: Partners*, 108–14, Center for Humans and Nature.

Lévi-Strauss, C. (1965), 'The Future of Kinship Studies', *Proceedings of the Royal Anthropological Institute* 1965:13–22.

Lyons, B. (1999), '"Taita Chimborazo and Mama Tungurahua": A Quichua Song, a Fieldwork Story', *Anthropology and Humanism* 24 (1): 32–46.

Madera, L. (2021), 'The Wondrous W'aka World: Lessons from Pach Mama in the Face of Despair', in G. Van Horn, R. Wall Kimmerer and J. Hausdoerffer (eds), *Kinship, Belonging in a World of Relations, Vol. 3: Partners*, 19–30, Center for Humans and Nature.

Malas, O. and M. D. Tolsá (2024), 'The Impact of Volcano Eruption on Mental Health: A Systematic Review', *International Journal of Disaster Risk Reduction* 113:104863.

Manning, E. (2019), 'Toward a Politics of Immediation', *Frontiers in Sociology* 3:42.

Massumi, B. (1995), 'The Autonomy of Affect', *Cultural Critique* 31:83–109.

Moreno, H., J. Naranjo and E. Polanco (2023), 'Geología del Complejo Volcánico Lonquimay-Tolguaca. Regiones del BioBío y de La Araucanía. Escala 1:50.000', Regiones del BioBío y de La Araucanía, SERNAGEOMIN.

Morim de Lima, A. and N. Soares-Pinto (2024), 'Kinship with the Land and Indigenous Cosmopolitics of Care', *Estudos Avançados* 38:173–94.

O'Grady, N. (2018), 'Geographies of Affect', in *Oxford Bibliographies*. https://doi.org/10.1093/obo/9780199874002-0186.

O'Grady, N. and D. Shaw (2023), 'Disaster Reparations? Rethinking Disaster Recovery Through the Politics of Affect', *Geographical Journal* 189 (3): 514–25.

Oracion, E. (2015), 'Kinship Networks and Resiliency to Flooding of Pagatban Riverside Communities in Negros Oriental', *Philippine Sociological Review* 63:27–51.

Paredes, J. P. (2013), 'El Presente Potencial y la Conciencia Histórica: Realidad Social, Sujeto y Proyecto. A la Memoria de Hugo Zemelman Merino', *Polis Revista Latinoamericana* 12 (36): 243–61.

Parra, E. (2022), 'Estrategias de Mujeres Frente al Riesgo de Desastres Por Peligros Volcánicos en Baños, Ecuador 1999–2020', Master's thesis, FLACSO, Quito, Ecuador.

Petit-Breuilh, M. (2004), *La historia eruptiva de los volcanes hispanoamericanos (siglos XVI al XX)*, Servicio de Publicaciones del Exmo, Cabildo Insular de Lanzarote-Casa de los volcanes.

Petit-Breuilh, M. (2016), 'Volcanes fronterizos en América Latina y la importancia de los comités de frontera en casos de desastre: Chile y Argentina en el siglo XX', in L. A. Arrioja and A. Alberola (eds), *Clima, desastres y convulsiones sociales en España e Hispanoamérica, siglos XVII–XX*, 324–45, Universidad de Alicante.

Polanco, E. (2010), 'Volcanoestratigrafía, geoquímica y peligro volcánico del volcán Lonquimay (38°30′S), Andes del Sur (Chile)', Doctoral diss., Universidad de Barcelona, Facultad de Geología, Departamento de Geoquímica, Petrología y Prospección Geológica.

Ramírez, V., L. A. Zambrano, M. C. Gutiérrez, A. Carvajal and T. Armijos (2017), 'Treinta años después de la erupción del volcán Nevado del Ruiz: memorias, voces, reparación y escenarios de participación', *Revista Colombiana de Sociales* 40 (1): 45–64.

Romero, J. E., F. Cáceres, L. Rojas, A. Guevara, E. Polanco, F. Vergara-Pinto, E. Michailidou, J. A. Naranjo, A. Fabbrizio, G. Weber, J. Bustillos, A. Castruccio, M. Poblete and I. Rodríguez (2025), 'Middle-to-Late Holocene Explosive Eruptions of Lonquimay Volcano (Southern Andes of Chile): Stratigraphy, Tephra Dispersal, and Pre-eruptive Conditions', *Journal of Volcanology and Geothermal Research* 468:108457.

Sahlins, M. (2011), 'What Kinship Is (Part One)', *Journal of the Royal Anthropological Institute* 17 (1): 2–19.

Salomon, F. (1986), *Native Lords of Quito in the Age of the Incas: The Political Economy of North Andean Chiefdoms*, Cambridge University Press.

Sandoval, V., C. Gonzalez-Muzzio and C. Albornoz (2014), 'Resilience and Environmental Justice: Potential Linkages', *Procedia Economics and Finance* 18:416–24.

Sarmiento, F., R. Chávez, C. Aguirre and J. Abrams (2022), 'Desarrollo sustentable y regenerativo de los paisajes socioecológicos de montaña: Montología del Chimborazo como referente insigne del cambio global', *Antropologías del sur* 9 (17): 121–45.

Seigworth, G. J. and M. Gregg (2010), 'An Inventory of Shimmers', in *The Affect Theory Reader*, 1–25, Duke University Press.

Setiawan, I., A. Tallapessy and A. Subaharianto (2022), 'Mount Bromo Will Take Care of Us: Tenggerese Religious-Ecological Knowledge, Challenge of Modern Reason, and Disaster Mitigation in Postcolonial Times', *Southeast Asian Studies* 11 (3): 399–426.

Sheets, P. D. and D. K. Grayson (eds) (1979), *Volcanic Activity and Human Ecology*, Elsevier.

Sigurdsson, H., S. Cashdollar and S. R. Sparks (1982), 'The Eruption of Vesuvius in AD 79: Reconstruction from Historical and Volcanological Evidence', *American Journal of Archaeology* 86 (1): 39–51.

Thrift, N. (2008), *Non-Representational Theory: Space, Politics, Affect*, Routledge.

Troncoso, L. (2022), 'El podcast como estrategia para difundir el conocimiento comunitario sobre peligros volcánicos: una aproximación desde métodos biográficos', Master's thesis, Universidad Andina Simón Bolívar/Corporación Editora Nacional, Quito, EC.

Troncoso, L. and A. Ordóñez (2023), 'Contra el mal del olvido: relatos de convivencia con un volcán en erupción', in P. Masías and M. Ortega (eds), *Libro de Resúmenes IX Foro Internacional de Peligros Volcánicos – IX FIPVO*, 350–4, Instituto Geológico Minero y Metalúrgico.

Vergara-Pinto, F. and A. Marín (2023), 'Stratigraphy of Volcanic Memory: Sociocultural Dimensions of Volcanic Risk in the Southern Andes, Chile', *Journal of Contingencies and Crisis Management* 31 (4): 1018–33.

Vergara-Pinto, F., N. O'Grady, A. Fredriksen, J. E. Romero, C. Marchant, R. Walshe, A. Donovan, J. Morin and M. Szlam (2024), 'How Residents and Volcanoes Co-Produce Risk Knowledge: Ways of Knowing and Affective Attunement to the Rhythms of Lonquimay volcano, Chile', *Journal of Volcanology and Geothermal Research* 454:108180.

Viveiros de Castro, E. (2009), 'The Gift and the Given: Three Nano-Essays on Kinship and Magic', in S. Bamford and J. Leach (eds), *Kinship and Beyond: The Genealogical Model Reconsidered*, 237–68, Berghahn.

Wächtershäuser, G. (2006), 'From Volcanic Origins of Chemoautotrophic Life to Bacteria, Archaea and Eukarya', *Philosophical Transactions of the Royal Society B: Biological Sciences* 361 (1474): 1787–808.

Weeks, S. and V. Popinsky (2016), 'Symbolism and Resilience in the Aftermath of a Destructive Volcanic Eruption', *Anthropology Now* 8 (2): 57–68.

Zemelman, H. (2010), 'Sujeto y subjetividad: la problemática de las alternativas como construcción posible', *Polis Revista Latinoamericana* 9 (27): 355–66.

8

Geographies of Governance

Disaster Response, Territorial Politics and the American Samoan Tsunami

Elissa Waters

Small island states are often positioned both as geographically marginal and at the centre of global concern about the increased frequency and severity of extreme events associated with climate change. In this context, small island developing states are portrayed as especially vulnerable to disaster due to their exposure to environmental hazards and their economic precarity. This framing, while not inaccurate, tends to rely on narrowly conceived indicators of risk and vulnerability and, in doing so, often obscures the nuanced and dynamic ways governance functions in small island contexts (Barnett and Waters 2016). In many cases, the governance systems that underpin disaster response are layered, historically informed, locally embedded and can be highly adaptable in times of crisis. Recognizing this complexity provides opportunities to challenge deficit-based assumptions about small island state capacity and opens up new ways of understanding disaster governance in context.

This chapter contributes to efforts within geography to situate disaster governance in its socio-political and spatial contexts (Tierney 2012; Cretney 2017). While disaster governance is often approached through institutional or policy analysis, such accounts can overlook the role of political history and cultural institutions in shaping governance outcomes (Hilhorst et al. 2020). They also tend to treat governance as a technical and linear process (Tierney 2012) rather than one mediated by long-standing relationships of power, authority and tradition. Drawing on insights from political geography, this chapter adopts a place-based approach to disaster governance, attending to the ways that scale, sovereignty and cultural institutions interact in complex and evolving ways.

Disasters offer a distinctive analytical lens through which to examine governance, precisely because they tend to expose otherwise obscured relationships and processes. In the Pacific region, and particularly in small island states and territories, disasters bring to light long-standing tensions between local leadership, external authority and the legacies of colonialism. This chapter uses the 2009 tsunami in American Samoa as a case through which to explore these dynamics. It argues that understanding disaster governance in American Samoa requires a careful reading of both its historical relationship with the United States and the cultural authority embedded within its village institutions.

In doing so, the chapter makes two key contributions. First, it demonstrates how disaster governance in American Samoa is shaped by a multi-scalar and relational understanding of sovereignty, in which federal and territorial powers operate in parallel, often overlapping and occasionally in tension. Second, it highlights the enduring role of cultural governance institutions, particularly the village and its associated social structures, in shaping effective disaster response. These findings suggest that disaster governance in American Samoa cannot be adequately understood through frameworks that privilege formal state institutions alone.

The analysis is grounded in fieldwork conducted in American Samoa in 2015, including interviews with community leaders, government officials and civil society actors, as well as archival and policy analysis. It begins by elaborating on the importance of geographical perspectives in understanding disaster governance before introducing the context of American Samoa as a US territory with a distinctive political and cultural structure. It then outlines the 2009 tsunami and the governance response, followed by an analysis of the multi-scalar governance dynamics of the United States–territorial relations and the role of traditional village institutions in disaster response. It concludes by reflecting on what the American Samoan case contributes to broader debates on disaster geographies, especially concerning scale, cultural politics and the plural configurations of authority in disaster response.

Disasters as a 'Window' into Place-Based Governance

Disasters are both shaped by governance systems and revelatory of them. Critical scholarship on 'natural' hazards has long argued that disasters stem less from the hazard itself and more from underlying structural factors within a society's political economy and social support systems (Albala-Bertrand 1993; Blaikie et al. 1994; Alexander 2000; Cutter et al. 2003). Empirical studies have built on this perspective to examine how disasters reshape relationships between the state, civil society and markets during and after crises. These interactions often involve coordination (or conflict) over aid distribution, reconstruction funding and the implementation of recovery programmes, where governments may rely on NGOs or private actors to fill gaps in service delivery, or where civil society groups push back against exclusionary or top-down interventions (Jalali 2002; Shaw and Goda 2004; Özerdem and Jacoby 2006; Schiff 2013; Srinivasan and Nagaraj 2013). In the context of these complex relations, disasters provide a unique lens for analysing governance, as they are inherently political and require interactions among a broad range of actors and institutions to mitigate their societal impact (Pelling and Dill 2008, 2010).

To understand how disasters can also reveal governance in action, it is useful to consider two dominant strands in the literature: vulnerability and disaster management. These are often treated as separate domains of inquiry, each providing a distinct analytical frame. Vulnerability approaches typically examine how long-standing social, political and economic processes shape uneven exposure to disaster risk, foregrounding questions of inequality, marginalization and historical injustice (Blaikie et al. 2014). Disaster management strategies, by contrast, tend to emphasize the technical and institutional arrangements for responding to and mitigating risk. These strategies are often future-oriented, drawing on risk projections and contingency models to guide intervention and are spatially organized across multiple institutions and scales (O'Grady 2018; Collier and Lakoff 2021). These frameworks remain vital but do not fully account for the ways disasters themselves can illuminate the dynamics of governance in place.

An alternative and complementary approach is to consider disasters not simply as the primary object of inquiry but as an analytical lens. Conventionally, disasters are treated as discrete events to be explained, triggering a search for causes, failures or responses within a risk management framework. In this mode, disasters are often approached through pre-existing analytical categories that are applied to assess governance impacts or interventions. By contrast, viewing disasters as a lens or point of entry for analysis foregrounds the ways in which they expose the workings of governance itself. Disasters can render visible the relationships, contradictions and gaps that exist within systems of authority, responsibility and institutional practice. In this sense, they 'speak back' to governance theory (for both disasters and broader political theory; see Waters 2019), revealing its limits, logics and fault lines in ways that are not always observable under ordinary conditions.

From an empirical point of view, the context of disasters also provides access to a wide and diverse governance landscape, enabling researchers to engage with a broader range of actors than might otherwise be available in more programmatic or sector-specific governance studies. In this sense, disasters can serve as valuable entry points for understanding the practical, relational and discursive dimensions of governance. Methodologically, they allow researchers to observe governance as it is enacted under pressure and to trace how authority, coordination and legitimacy are negotiated in real time.

This chapter draws on a mixed-methods qualitative study conducted over five months in American Samoa in 2015. The research design centred on an interpretive, place-based approach that sought to understand how governance was experienced, described and enacted by those involved in the 2009 tsunami response and recovery. Fieldwork included fifty-one semi-structured interviews with government officials, traditional leaders, civil society actors and community members; participant observation embedded within the Territorial Emergency Management Coordinating Office; archival research into historical and legislative documents; and a media analysis of news reporting in the year following the tsunami. A central aim of the research was to foreground the ideas, experiences and concerns of those who construct governance in place every day. Throughout the chapter, extended quotations are used not simply as illustrative examples but as a core method of analysis – offering insights into how governance is narrated, negotiated and contested by different actors. By allowing space for these perspectives to be voiced in their own terms, the chapter seeks to reveal the diverse and sometimes contradictory ways in which disaster governance is understood and enacted in American Samoa.

Importantly, interviewees generally viewed the disaster governance positively – an uncommon perspective in the existing literature on post-disaster response (Malhotra and Kuo 2007). While many studies emphasize breakdowns, failures or external imposition, the American Samoan case offers a more situated and often affirmative account. This adds important nuance to debates on what constitutes 'good governance' in disasters, particularly in contexts where legitimacy is built through local cultural practices and relational political authority.

Situating American Samoa

American Samoa is an unincorporated and unorganized territory of the United States located in the South Pacific. Comprising five volcanic islands and two atolls, it is home to a population of just over 55,000. While geographically distant from the US mainland, American Samoa is deeply connected to it through legal, political and financial systems. These ties shape not only its day-to-day governance but also the ways in which disaster response and recovery are organized and experienced.

The political status of American Samoa as an unincorporated territory is distinctive. Unlike incorporated US territories, the US Constitution does not fully apply in American Samoa, and residents are US nationals rather than citizens. Its relationship with the United States was formalized through the deeds of cession in 1900 and 1904, in which local chiefs agreed to US sovereignty while preserving key aspects of customary governance. This has produced a hybrid system in which traditional leadership structures – especially the matai chiefly system – coexist with formal state institutions under the broader umbrella of US territorial governance. The territory has its own constitution, legislature and judiciary, but authority is layered, with local decision-making operating within the broader political and financial oversight of the United States.

Crucially, governance in American Samoa cannot be understood through a strictly institutional lens. Traditional village-based authority remains central to political life. Chiefs (matai) hold leadership roles not only within extended families and villages but also in the territory's formal political institutions. Many government officials are also senior traditional leaders, and key decisions around land, leadership and community organization continue to be shaped by customary norms. The village remains the primary site of political identity and practice, and cultural values continue to underpin local governance. These cultural systems are not parallel to state governance but are entangled with it, often shaping how authority is understood, negotiated and exercised in everyday practice.

This institutional and cultural hybridity becomes especially visible during disasters. In crisis situations, local authorities and village leaders play central roles in immediate response while external support – particularly from the Federal Emergency Management Agency (FEMA) and other US agencies – provides vital financial and logistical resources. The result is a governance landscape defined by shared, negotiated and sometimes contested authority, where formal sovereignty, cultural legitimacy and federal dependence intersect.

Understanding this context is essential for interpreting disaster governance in American Samoa. The response to the 2009 tsunami cannot be explained solely by state capacity or policy design. Rather, it reflects the layered and relational nature of governance in a US territory shaped by colonial legacies, cultural traditions and multi-scalar political ties. The following section introduces the tsunami and its aftermath, exploring how these dynamics unfolded in practice.

The 2009 Tsunami Response

At 6.48 am on 29 September 2009, an 8.3 magnitude earthquake – the largest of the year worldwide – struck 196 kilometres south-west of American Samoa, triggering a devastating tsunami. Within fifteen minutes, waves inundated the coasts of American Samoa, Samoa and Tonga, resulting in the deaths of 189 people and causing widespread destruction. Samoa bore the brunt of the disaster, with 146 fatalities and the near-total devastation of villages in south-eastern Upolu. In Tonga, nine lives were lost, and significant structural damage was reported.

In American Samoa, an incredibly timed, pre-scheduled national tsunami drill contributed to a reduction in potential casualties, as authorities were ready to respond, having undergone recent training. Nevertheless, the speed of the impacts left minimal time for formal emergency warnings. As a result, evacuations were extremely challenging; thirty-four people lost their lives, with two bodies never recovered. The economic centre of the island, Pago Pago, experienced severe damage, and many other parts of the island were also affected, leading to the displacement of approximately 2,200 residents and the destruction of 248 homes and a school. Infrastructure repairs, including power plant restoration, incurred costs amounting to $21.8 million, while additional damage to water and wastewater systems further complicated recovery efforts.

Governing the Response

The timeline of the 2009 tsunami response (Figure 8.1) highlights the complexity of governance in the American Samoan disaster. This section draws on that sequence to examine how the response unfolded in practice and what it reveals about the layered and relational nature of disaster governance in the territory.

The immediate response to the tsunami was heavily reliant on village institutions, revealing their centrality in disaster preparedness and early action, with local leaders playing a decisive role in issuing evacuation warnings and

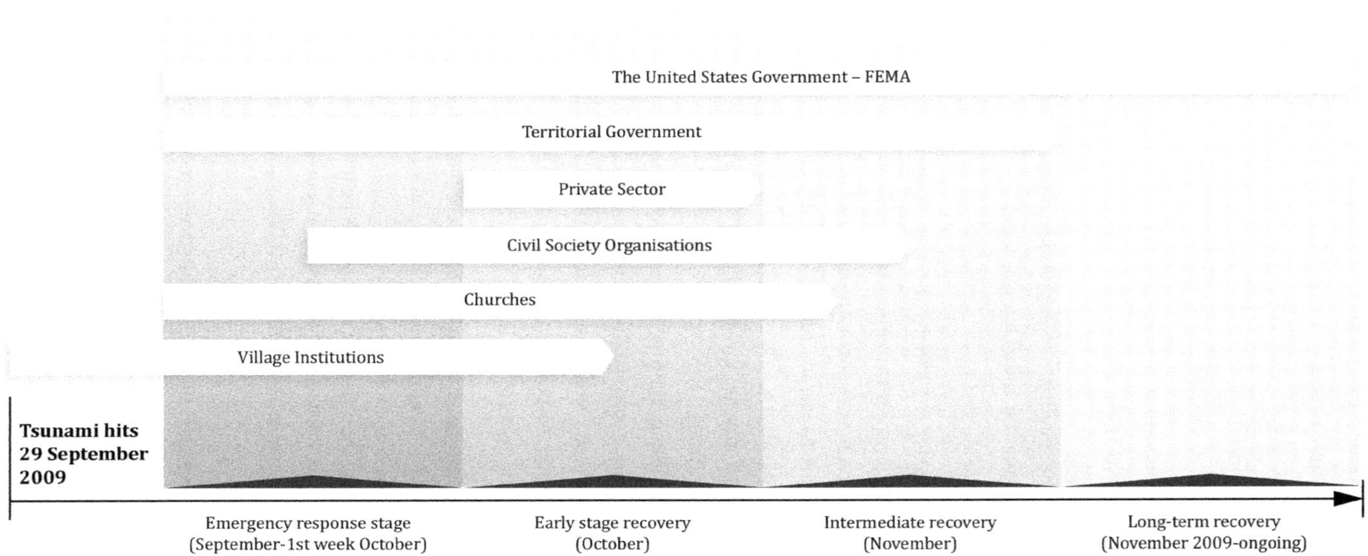

Figure 8.1 The main governance actors and institutions involved in the disaster response and recovery.
Source: Author.

coordinating relief (Binder et al. 2014; Rumbach and Foley 2014). While local leadership facilitated a swift community response, government agencies encountered difficulties with coordination.

The Emergency Operations Centre (EOC) faced communication failures that delayed the deployment of emergency medical teams.

> It was just chaos, because nobody had ever even nearly expected that kind of incident. It felt like everybody in the EOC was in chaos. . . . We realised we needed to be the people in charge, so we needed to slow down and calm down and get a handle on the situation. . . . Then we just followed the plan. I would say the plan worked well after the initial hour of chaos.
>
> – Government employee

Following the initial disaster, the American Samoan government quickly initiated the federal disaster declaration process, which led to an unusually prompt US intervention. President Obama formally declared a major disaster that same evening, allowing FEMA to deploy an Incident Management Assistance Team to coordinate relief operations. The Department of Health, in collaboration with FEMA, distributed emergency supplies while the US military facilitated essential logistical operations.

Although initial coordination proved difficult, relief efforts became increasingly structured as roles and responsibilities were defined. The arrival of non-governmental organizations (NGOs) signalled a transition into a more politically complex phase of disaster response, and some challenges arose in reconciling federal intervention with existing territorial governance structures.

> There were so many people in charge - They were like "What? A tsunami hit American Samoa? We're flying!!" (laughs) So, like in all disasters here there was a lot of confusion. The reason that happens is because there are way too many people who want to call the shots and not enough people to do the work. So that creates chaos and it's not needed, and that's what happened for a time in 2009 – I don't know if people are going to admit to that but, yeah, it happened.
>
> – Government employee

To navigate these complexities, coordination meetings between FEMA representatives, the American Samoan government and village chiefs became instrumental in ensuring the culturally sensitive and equitable distribution of aid.

The American Red Cross and local NGOs played a key role in bridging gaps between federal agencies and affected communities. Their involvement ensured that aid distribution adhered to local customs and that relief efforts were tailored to community needs. Despite early logistical hurdles, the interviewees were adamant that sustained collaboration between government agencies, NGOs and village leaders ultimately reinforced disaster governance and improved overall response effectiveness in the initial stages of recovery.

The long-term recovery of American Samoa was significantly dependent on financial assistance from the US government. In total, FEMA allocated $22.4 million in individual assistance, while the US Small Business Administration provided $11.4 million in low-interest loans to affected businesses. Public assistance funding totalled $57 million, supplemented by an additional $25 million to sustain ongoing recovery projects. Although logistical and bureaucratic challenges arose during implementation, the overall response was widely regarded as effective in supporting survivors and aiding economic reconstruction.

The response to the 2009 tsunami was a multifaceted and dynamic process, marked by a complex interplay between local leadership, federal intervention and NGO participation. Village institutions played a crucial role in immediate evacuations, while US federal agencies provided essential financial and logistical support. The disaster underscored the resilience of Samoan communities and highlighted the need for disaster governance systems that effectively balance external intervention with local leadership structures.

This analysis highlights two key points that form the foundation of a detailed examination of place-based disaster governance relations. First, it underscores the critical role of the US government in territorial governance, as the effectiveness of the American Samoan government's response depended heavily on this relationship. Second, it emphasizes the village as a central space for organization and response, while also showing how governance roles extend beyond village boundaries. The following sections demonstrate how geographical approaches – particularly those attuned to history, place and scale – can illuminate the politics of disaster governance and unsettle dominant assumptions about how power, responsibility and legitimacy are organized in times of crisis.

Extraterritorial Governance and Complexities of Sovereignty

Disasters such as the tsunami in American Samoa expose the complexities of sovereignty and governance beyond territorial boundaries. They reveal the intricate bureaucratic flows and power dynamics that transcend conventional understandings of territory and political authority (Allen 2016). This is particularly relevant in the Pacific, where colonial legacies continue to shape governance in small island states.

Understanding the effectiveness of American Samoa's disaster response requires situating it within the region's historical and political context. Pacific small island states and subnational island jurisdictions (SNIJs) have long been influenced by external forces including colonialism, trade, aid and tourism (Baldacchino 2012). Many remain economically reliant on external connections (Richards 1982) while larger states have historically viewed them as strategic or economic enclaves (Mountz 2015). Today, fifty-five island territories remain autonomous yet non-sovereign, maintaining varying degrees of political and economic dependence within wider sovereign state spaces (Ferdinand et al. 2020).

Governance in these territories is shaped by long histories of colonial and imperial control (Baldacchino 2004). Some remain under direct metropolitan rule while others operate under semi-autonomous arrangements that enable them to negotiate political and economic advantages. SNIJs in particular challenge traditional notions of sovereignty, demonstrating how smaller entities can leverage dependencies to influence larger powers (Neumann and Gstöhl 2004; Baldacchino and Milne 2006).

Political geographers analyse these governance dynamics, questioning the conventional link between territory, sovereignty and authority (Clapham 1998; Agnew 2005; Hansen and Stepputat 2009; McConnell 2009), and highlighting how island geographies often foster innovative governance models that blend external and local structures (Baldacchino and Milne 2006; Stratford 2006; Grydehøj 2016). This dynamic is particularly evident in American Samoa, where disaster response mechanisms are shaped by both historical agreements and contemporary political arrangements.

American Samoa and the United States: A Mutually Beneficial Relationship

The relationship between the United States and the eastern islands of Samoa emerged during the broader wave of colonial expansion into the Pacific in the nineteenth century, reflecting a unique case of strategic alignment amid imperial competition. However, unlike many colonial encounters characterized by one-sided exploitation, this relationship was shaped by mutual strategic interests rather than economic extraction. While the US Navy valued the islands for their military and geostrategic utility, the Samoan political elite also recognized the benefits of aligning with a powerful metropolitan state (Sunia 1998; Shaffer 2000). Amid ongoing regional instability, Samoan high chiefs saw US protection as a means to secure political stability, reinforce their status within the trans-Samoan chiefly system and preserve their authority over land, culture and governance (Armstrong 2009). Although the balance of power undoubtedly favoured the United States, Samoan leaders were not merely passive participants in the cession process but actively engaged in shaping the terms of their relationship with the United States.

In 1900, the chiefs of Tutuila formally ceded control of the islands to the United States through the Articles of Cession, a document that continues to define American Samoa's relationship with the United States. The chiefs' decision to cede sovereignty was strategic, aiming to ensure political stability while safeguarding their control over land and titles, which were central to their legitimacy. For its part, the United States required a locally endorsed cession to remain consistent with its ideological opposition to formal colonialism and its perceived responsibility to preserve the Samoan cultural system (Laughlin 1982).

The transition from naval rule to the current status of an 'unorganised' and 'unincorporated' territory began in the 1960s with the drafting of a territorial constitution. This process continued over the following decade through careful negotiations involving both the United States government and the emerging territorial government (Sunia 1998). The institutions and governance mechanisms established – such as the legislature and constitution – reflect a careful balance between territorial autonomy and metropolitan oversight. They aim to balance the terms of the deeds of cession, the principles of the US Constitution and the practicalities of governing a transitioning territory (Laughlin 1980; Morrison 2013).

Today, the Articles of Cession and the commitments outlined in the constitution remain foundational to American Samoa's political identity and continue to shape perceptions of US obligations to the territory. While not formally recognized as a treaty, many American Samoans refer to the agreement as such, reinforcing the prevailing belief that the United States remains accountable for its commitments to protection, land rights and self-governance. This enduring arrangement continues to influence how American Samoans negotiate questions of autonomy, identity and obligation within the US territorial framework.

Disaster Response and Relative Authority: The Case of FEMA

The response to the tsunami illustrates the complex and layered authority shared between the US federal government, represented by FEMA, and the American Samoan government. While FEMA is legally responsible for coordinating federal responses, it does not override local decision-making – mirroring the federal-state dynamics found within the continental United States. This approach seeks to balance pragmatism with effective disaster management, reflecting broader dynamics in the United States–American Samoa relationship.

Many interviewees confirmed that the territorial government leads disaster response, with FEMA in a supporting role:

> We were in charge. When any federal government partners come on ground, the local government is always in charge. They are a support role. So, we tell them we need this, that, and that. And within a day and a half, we had FedEx landing with all these medical supplies.
>
> – Government employee

However, some officials described FEMA as taking on a managerial role in practice:

> So, when they come down here, they pretty much just ask me where things are and some statistics, and I pretty much let them . . . well, I don't let them . . . they take over the management of things, you know, the distribution of equipment and things. Well, first of all, it's their stuff, you know? (laughs) But we help out to suggest, this should go here and this there.
>
> So, who has ultimate authority there? Is that you or FEMA? (Interviewer)
>
> Well, basically, FEMA has the authority over the distribution of funds, but I have the authority to designate locally where those funds and equipment should go. And they have professionals that go and check on that.

This distinction between de jure (legal) authority and de facto (practical) authority highlights the perceived value of the financial and technical expertise FEMA brings to disaster response. While American Samoa retains formal authority, it frequently relies on FEMA's resources and expertise in practice. Interestingly from a cultural perspective, interviewees often framed this relationship in familial terms:

> So, we're like a little nephew out there in the middle of nowhere, and it has been that way ever since 1900. We might be an annoying little nephew sometimes, but you have to help your family (laughs).
>
> – Traditional leader

The relationship between the United States and American Samoa has evolved over time and is shaped by both strategic considerations and mutual interests. While the Articles of Cession continue to define American Samoa's political framework, the dynamics of governance – in both day-to-day political administration and disaster response – demonstrate a balance between local leadership and US support. Despite being an unincorporated territory, American Samoa maintains a distinct identity, with governance structures reflecting both local mechanisms and US legal frameworks. The ongoing engagement between territorial authorities and federal agencies such as FEMA underscores the complexity of this unique political relationship and underpins a large part of the effective governance response to the 2009 tsunami.

The following section explores the second key influence: the role of traditional governance institutions and their evolving relationship with the territorial state.

Cultural Governance and the Complexities of the Local State

Recognizing the role of local cultural governance institutions in disaster response is especially critical in small island states like American Samoa, where governance systems are shaped by complex interactions between traditional leadership and the local state. While much of the disaster governance literature is focused on localization efforts aimed at shifting disaster management to local actors, Melis and Apthorpe (2020) argue that this approach often overlooks multi-local governance complexities. Small island states, with their complex sovereignty arrangements and layered forms of authority, exemplify the need for a more nuanced understanding of local governance.

In many island contexts, including American Samoa, traditional leadership structures and religious organizations play a central role in disaster governance yet are often perceived as static rather than adaptive institutions (Maldonado 2016). Manyena (2014), for example, highlights how African chieftaincies maintain governance functions by adapting to socio-political and environmental changes – a pattern also evident in island nations such as American Samoa, where customary governance remains deeply embedded in local decision-making practices. Religious institutions, also historically overlooked in disaster management, contribute significantly to psychological support, community mobilization and disaster relief (Ha 2015). Recognizing these institutions as dynamic and integral to governance is essential to understanding how cultural governance functions at the village level in American Samoa and how it intersects with the authority of the local state in shaping disaster response.

The Primacy of the Village as Political Space in American Samoa

As a political space, the village has a far longer history than the 'national' sphere of modern government in the Samoan islands. Traditionally, villages in American Samoa were politically autonomous, governed by matai (chiefs) within a hierarchical system of titled rights to customary land. Over time, however, this authority has evolved – particularly with the arrival of missionaries and the US government – leading to a complex system of traditional authority that remains grounded in village institutions while extending across spatial boundaries.

Traditionally, a *nu'u* (village) functioned as a self-governing entity, with chiefs from *āiga* (extended families) holding ultimate authority over day-to-day governance (Meleisea 1987). Each village had a *fa'avae* (constitution) that set out governance rules, including the *fa'alupega* – an honorific code ranking families

Indigenous Socio-political Institutions	Roles and Responsibilities
Matai - Holders of chiefly titles	Allocates and manages family land, distributes village resources and assets, resolves disputes, represents family interests in councils
Aiga - Extended families	Ensures wellbeing of other members of the *'aiga*, serves the *matai* through financial and in-kind contributions
Aualuma - Organization of women; women's group	Provides hospitality to village guests, village presentation, cleanliness and sanitation and health
Aumaga - Organization of untitled men	Maintaining the village grounds and infrastructure; heavy labour; fishing and planting crops, enforcing village regulations, maintaining safety

Figure 8.2 Samoan institutions and roles and responsibilities.
Source: Adapted from Rumbach and Foley (2014); Huffer and So'o (2005); Shore (1982) and the Revised American Samoa Constitution (1967).

and chiefly titles. This structured hierarchy was reinforced through the village *fono* (council), which regulated everything from land disputes to social customs (Amosa 2010).

Today, governance at all levels remains deeply embedded in the *fa'asamoa*, the foundational socio-political system of both Samoa and American Samoa. Consisting of cultural practices, shared symbols and agreed principles, it underpins nearly every aspect of daily life – social, political and economic (Le Tagaloa 1992). Importantly, it also defines the roles of individuals and groups in relation to one another (Gershon 2006). Figure 8.2 outlines the main institutions and roles that operate in Samoan villages.

A fundamental principle that underpins these roles and responsibilities is *tautua* (service), which binds matai to their *āiga* through mutual accountability. Traditionally, service was most often provided through labour and active participation in village institutions, as well as through ceremonial contributions to head matai in the form of food and *'n ttoga* (ceremonial fine mats) during times of need known as *fa'alavelave* – a particularly Samoan concept described in the following exchange:

> Can you explain fa'alavelave for me? (Interviewer)
>
> I don't want to say a disaster, but it is kind of a disaster. So, like a funeral everyone comes together. But it's not always a crisis, weddings are fa'alavelave too. So, it's anything that would cause a lot of burden on one individual, where instead people come together.
>
> – Government employee

This communal approach to governing hardship contrasts with the individualistic models in capitalist democracies, as a paramount chief noted in this telling explanation:

> So, you're a palagi (Caucasian foreigner) right, if I give you $1000 what are you going to do with it? Probably go buy a new phone, pay off your credit card, it's all about taking care of you. If you give me $1000, I'll go give some to the pastors first, ok? Then I'll give some to my wife, make sure she gives some to my immediate family. But I'll take most of it and keep it for my extended family, for fa'alavelave. Then of course if there is a government function I'll give to that. Now here's the catch, if I have a fa'alavelave, my family will give it back, and the pastor will pray for me. In your case, you just have yourself. It's like a machine, or a recycle system. It goes out and comes back in.
>
> – Traditional leader

This structured reciprocity is fundamental to village life and is tied to the right to reside in a village, as another interviewee explained:

> That is fa'alavelave. If you don't (pay service), after a while the matai can evict you. You can't just go and live there and shut your doors and say don't bother me I'm watching TV. No, no, no, it's a communal lifestyle.
>
> —Government employee

Importantly, from a governance perspective, this responsibility of service is not typically considered at a national or island-wide scale but rather remains confined to the āiga and the village. These village-based systems of authority continue to structure how governance is experienced and practised at the grassroots level in American Samoa.

Cultural Governance during the Tsunami

During the 2009 disaster response, established village hierarchies and the norms and practices that support their roles proved critical. The *pulenu'u*, *aumaga* and *aualuma* were essential first responders, acting swiftly and effectively within their established roles and responsibilities. As one interviewee described:

> Islander people work in teams. Culturally and traditionally, they work in teams. Those teams are functional groups that are already in place, and they are disciplined groups. So, this is how people will function and respond in disasters. Each group with its own responsibilities will sweep through the village doing what they need to do and they will meet up with the other groups knowing exactly what those other groups are doing as well. They will have command posts but not like you would think of a command post - mostly they'll be standing under a breadfruit tree, or what's left of it, and they don't need to be referring to a plan at that point.
>
> - Government employee

When asked about their role in disaster governance, most of the senior matai interviewees indicated that their main responsibility was to ensure the organization of the village structure, as this interviewee explained:

> I just needed to tell the village mayors 'be aware of what your functions are and do that, if you have a problem come to me and I'll sort it out' but each has a role, and I don't need to do it for them.
>
> - Traditional leader

The disaster response also confirmed the importance of familial connections in structuring governance practices within and beyond the village, and in the early stages of the tsunami, these were integral. This emphasis on kinship-based cooperation emerged repeatedly in discussions about local resilience and recovery. Some interviewees made comparisons with other parts of the world where collective familial support is not as strong, as in this discussion:

> We're thankful for the culture and having tight knit families because it makes us very resilient people. Yes, something horrible can happen but we will do what we need to do to overcome that, and that's all because we have that village system where we can lean on one another. It's not like the US mainland, and I lived there, you don't have that, there is a disconnect. You can't rely on your neighbours. They're not going to come and offer you the help like you can get here.
>
> – Government employee

There are parallels here to the familial language used to describe the relationship with the US government. As with so many aspects of American Samoan governance, the balance of relative authority across scales remains fluid and subject to change and represents just one element in the evolving constellation of practices that constitute American Samoa's governance regime.

Conclusion

The response to the 2009 tsunami in American Samoa highlights the complexity of disaster governance in small island territories, where authority is layered, relational and grounded in both political history and cultural practices. Rather than exposing institutional breakdowns, as is often the focus of disaster literature, the tsunami response revealed a functioning, adaptive and culturally embedded system of governance. Local leadership, traditional institutions and established relationships with external actors all contributed to a response that was widely viewed as effective. These dynamics challenge standardized narratives of disaster response and suggest the need for more contextually grounded understandings of governance in times of crisis.

This chapter has argued that disaster governance in American Samoa benefits greatly from being understood through a political geographic lens – one that takes seriously questions of scale, sovereignty and place. The territory's unincorporated status, its historical relationship with the United States and the enduring role of customary authority all combine to produce a governance landscape that resists easy categorization. As a US territory with a strong tradition of customary governance, American Samoa represents a case in which state authority is only one part of a broader, overlapping system. During disasters, these systems do not compete so much as cohere – often in improvisational but highly effective ways – drawing on long-standing cultural principles of service, reciprocity and community leadership.

These findings add nuance to ongoing debates about 'good governance' in disasters. In the American Samoan context, legitimacy and effectiveness are determined not by institutional capacity alone but also by the strength of social relationships and the embeddedness of cultural political authority. At the same time, the tsunami exposed a governance system that is at once deeply local and profoundly shaped by external ties. The territory's reliance on federal assistance raises important questions about the sustainability of such systems in the face of shifting political relationships, changing expectations of aid and the long-term influence of US disaster policy on local governance practices.

By taking a grounded approach to disaster governance, this chapter contributes to wider efforts to situate disaster response within specific social, political and spatial contexts. As other contributions in this volume suggest, disasters are tied to the deeper structures and relationships that underpin governance in place. In American Samoa, those structures are shaped by colonial legacies, a culturally embedded political system and a practical understanding of how to navigate authority across scales. Recognizing this complexity is crucial not only for how we study disasters but also for how we understand the everyday geographies of governance in small island places.

References

Agnew, J. (2005), 'Sovereignty Regimes: Territoriality and State Authority in Contemporary World Politics', *Annals of the Association of American Geographers* 95 (2): 437–61.

Albala-Bertrand, J. M. (1993), *The Political Economy of Large Natural Disasters*, Clarendon Press.

Alexander, D. (2000), *Confronting Catastrophe: New Perspectives on Natural Disasters*, Terra Publishing.

Allen, J. (2016), *Topologies of Power: Beyond Territory and Networks*, Routledge.

Amosa, D. U. (2010), 'Local Government and Good Governance: The Case of Samoa', *Commonwealth Journal of Local Governance* 7:7–21.

Armstrong, K. V. (2009), 'The Weight of Names in American Samoa', *Ethnology* 48 (1): 53–69.

Baldacchino, G. (2004), 'Autonomous But Not Sovereign? A Review of Island Sub-Nationalism', *Canadian Review of Studies in Nationalism* 31 (1–2): 77–89.

Baldacchino, G. (2012), 'The Lure of the Island: A Spatial Analysis of Power Relations', *Journal of Marine and Island Cultures* 1 (2): 55–62.

Baldacchino, G. and D. Milne (2006), 'Exploring Sub-national Island Jurisdictions: An Editorial Introduction', *Round Table* 95 (386): 487–502.

Barnett, J. and E. Waters (2016), 'Rethinking the Vulnerability of Small Island States: Climate Change and Development in the Pacific Islands', in J. Grugel and D. Hammett (eds), *The Palgrave Handbook of International Development*, 731–48, Palgrave Macmillan.

Binder, S., C. Baker and J. Mayer (2014), 'Resilience and Recovery in American Sāmoa: A Case Study of the 2009 South Pacific Tsunami', *Journal of Community Psychology* 42 (7): 799–822.

Blaikie, P., T. Cannon and I. Davis (1994), *At Risk: Natural Hazards, People's Vulnerability and Disasters*, Routledge.

Blaikie, P., T. Cannon, I. Davis and B. Wisner (2014), *At Risk: Natural Hazards, People's Vulnerability and Disasters*, Routledge.

Clapham, C. (1998), 'Degrees of Statehood', *Review of International Studies* 24:143–57.

Collier, S. J. and A. Lakoff (2021), *The Government of Emergency: Vital Systems, Expertise, and the Politics of Security*, Princeton University Press.

Cretney, R. M. (2017), 'Towards a Critical Geography of Disaster Recovery Politics: Perspectives on Crisis and Hope', *Geography Compass* 11 (1): e12302.

Cutter, S. L., B. J. Boruff and W. L. Shirley (2003), 'Social Vulnerability to Environmental Hazards', *Social Science Quarterly* 84 (2): 242–61.

Ferdinand, M., G. Oostindie and W. Veenendaal (2020), 'A Global Comparison of Non-Sovereign Island Territories: The Search for "True Equality"', *Island Studies Journal* 15 (1): 43–66.

Gershon, I. (2006), 'When Culture Is Not a System: Why Samoan Cultural Brokers Can Not Do Their Job', *Ethnos* 71 (4): 533–58.

Grydehøj, A. (2016), 'Navigating the Binaries of Island Independence and Dependence in Greenland: Decolonisation, Political Culture, and Strategic Services', *Political Geography* 55:102–12.

Ha, K. M. (2015), 'The Role of Religious Beliefs and Institutions in Disaster Management: A Case Study', *Religions* 6 (4): 1314–29.

Hansen, T. and F. Stepputat (2009), *Sovereign Bodies: Citizens, Migrants, and States in the Postcolonial World*, Princeton University Press.

Hilhorst, D., K. Boersma and E. Raju (2020), 'Research on Politics of Disaster Risk Governance: Where Are We Headed?', *Politics and Governance* 8 (4): 214–19.

Huffer, E. and A. So'o (2005), 'Beyond Governance in Sāmoa: Understanding Samoan Political Thought', *Contemporary Pacific* 17 (2): 311–33.

Jalali, R. (2002), 'Civil Society and the State: Turkey after the Earthquake', *Disasters* 26 (2): 120–39.

Laughlin, S. (1982), 'United States Government Policy and Social Stratification in American Samoa', *Oceania* 53 (1): 29–38.

Laughlin, S. K., Jr. (1980), 'The Application of the Constitution in United States Territories: American Samoa, a Case Study', *University of Hawaii Law Review* 2 (2): 337–88.

Le Tagaloa, A. F. A. (1992), 'The Samoan Culture and Government', in R. Crocombe (ed.), *Culture and Democracy in the South Pacific,* 117–37, Institute of Pacific Studies, The University of the South Pacific.

Maldonado, J. (2016), 'Considering Culture in Disaster Practice', *Annals of Anthropological Practice* 40 (1): 52–60.

Malhotra, N. and A. Kuo (2007), 'Attributing Blame: The Public's Response to Hurricane Katrina', *Journal of Politics* 70 (1): 1–16.

Manyena, S. B. (2014), 'Disaster Resilience: A Question of "Multiple Faces" and "Multiple Spaces"?', *International Journal of Disaster Risk Reduction* 8:1–9.

McConnell, F. (2009), 'Governments-in-Exile: Statehood, Statelessness and the Reconfiguration of Territory and Sovereignty', *Geography Compass* 3:1902–19.

Meleisea, M. (1987), *The Making of Modern Samoa: Traditional Authority and Colonial Administration in the History of Western Samoa*, editorips@usp.ac.fj.

Melis, S. and R. Apthorpe (2020), 'The Politics of the Multi-Local in Disaster Governance', *Politics and Governance* 8 (4): 366–74.

Morrison, S. (2013), 'Foreign in a Domestic Sense: American Samoa and the Last US Nationals', *Hastings Constitutional Law Quarterly* 41 (1): 71–149.

Mountz, A. (2015), 'Political Geography II: Islands and Archipelagos', *Progress in Human Geography* 39 (5): 636–46.

Neumann, I. B. and S. Gstöhl (2004), *Lilliputians in Gulliver's World: Small States in International Relations*, Centre for Small State Studies, University of Iceland.

O'Grady, N. (2018), *Governing Future Emergencies: Lived Relations to Risk in the UK Fire and Rescue Service*, Springer.

Özerdem, A. and T. Jocoby (2006), *Disaster Management and Civil Society: Earthquake Relief in Japan, Turkey and India*, I.B. Tauris.

Pelling, M. and K. Dill (2008), 'Disaster Politics: From Social Control to Human Security', Environment, Politics and Development Working Paper Series, 1–24, Department of Geography, King's College London.

Pelling, M. and K. Dill (2010), 'Disaster Politics: Tipping Points for Change in the Adaptation of Sociopolitical Regimes', *Progress in Human Geography* 34 (1): 21–37.

Revised American Samoa Constitution (1967), 'Revised Constitution of American Samoa (with Links to 2022 and 2025 Amendments)', 2 June. Available at: https://www.refworld.org/legal/legislation/natlegbod/1967/en/150222 (12 August 2025).

Richards, J. (1982), 'Politics in Small Independent Communities: Conflict or Consensus?', *Journal of Commonwealth & Comparative Politics* 20 (2): 155–71.

Rumbach, A. and D. Foley (2014), 'Indigenous Institutions and Their Role in Disaster Risk Reduction and Resilience: Evidence from the 2009 Tsunami in American Samoa', *Ecology and Society* 19 (1).

Schiff, J. S. (2013), 'Silencing the Opposition: The State v. Civil Society in India's Ganges River Basin', *International Studies Perspectives* 15 (2): 229–42.

Shaffer, R. (2000), *American Samoa: 100 Years Under the United States Flag*, Island Heritage Publishing.

Shanmugaratnam, N. (2005), 'The Spectre of a "Second Tsunami" in Sri Lanka: What Can We Do to Prevent a Human-Made Disaster?', *Polity* 2 (3): 12–14.

Shore, B. (1982), *Sala'Ilua a Samoan Mystery*, Columbia University Press.

Shaw, R. and K. Goda (2004), 'From Disaster to Sustainable Civil Society: The Kobe Experience', *Disasters* 28 (1): 16–40.

Srinivasan, K. and V. Nagaraj (2013), 'The State and Civil Society in Disaster Response', *Journal of Social Work in Disability and Rehabilitation* 5:57–80.

Stratford, E. (2006), 'Technologies of Agency and Performance: Tasmania Together and the Constitution of Harmonious Island Identity', *Geoforum* 37 (2): 273–86.

Sunia, F. I. F. (1998), *The Story of the Legislature of American Samoa*, The Legislature of American Samoa.

Tierney, K. (2012), 'Disaster Governance: Social, Political, and Economic Dimensions', *Annual Review of Environment and Resources* 37 (1): 341–63.

Waters, E. (2019), 'Placing Theories of Governance: A Political Geography of American Samoa', PhD thesis, University of Melbourne, Australia.

9

Urban Disaster Risk and Risk Reduction as Multi-Scalar Configurations

Theresa Zimmermann

This chapter argues that understanding configurations of urban disaster risk governance requires critical inquiry into spatially interlinked processes of framing risks and associated forms of knowledge and power, for which human geographical thought can provide valuable perspectives. Over the past decade, urban disasters and related efforts to reduce risks have received increasing attention from international policy circles as well as academia. Policy frameworks such as the Sendai Framework for Disaster Risk Reduction 2015–30 (SFDRR), the Transforming our World: 2030 Agenda for Sustainable Development (SDGs), the Paris Agreement on Climate Change and the New Urban Agenda (NUA) recognize the shift from disaster management and relief towards understanding and reducing risks and emphasize the importance of risk reduction at local governance levels. Irrespective of how the aforementioned agendas are implemented nationally and locally, they have immense potential to influence discourse and practice in science, politics, planning and society, for example, with regard to funding allocated for research and development cooperation. Several international networks, NGOs and initiatives have formed in the past decades to bridge global expectations and local needs to support cities in becoming resilient, to reduce disaster risks and to identify synergies and potential overlaps with adaptation to climate change impacts.[1] Further, various academic disciplines have examined questions of resilience, disaster governance, risk management and adaptation planning in, by and for cities. Many scholars argue that these policy and planning processes are normative and political, reflect power relations, can reproduce inequalities and result in maladaptive outcomes (Vale 2014; Anguelovski et al. 2016; Meerow and Mitchell 2017; Ziervogel et al. 2017; Grove 2018b; Fünfgeld and Schmid 2020; O'Grady and Shaw 2023). Previous research has examined how local risk and adaptation governance (e.g. municipal or neighbourhood levels) is linked to, embedded in or shaped by policy processes at other scales, for example, through processes of decentralization, multilevel governance or transnational networks (Fünfgeld 2015; Chu 2018). However, there is less insight into how relations of power and knowledge at and between various scales shape the constitution of an urban risk governance regime and associated governing techniques used to address or attend to disasters. The choice of risk-governing techniques is facilitated by processes through which urban risks are problematized through particular forms of knowledge and underlying rationalities and thereby turned into objects of governance.

Against this background, this chapter draws on human geographic scale theory and Foucauldian thought to study urban disaster risk governance and multi-scalar processes of knowledge production more broadly. Building on the notion of regimes of practice from the field of governmentality studies, the chapter assesses how risk reduction in, for and through the urban is configured within a web of different ways to problematize flood risks and associated techniques developed and implemented at local to global scales to govern such risks. I use scale and the politics of scale to assess how different prevalent problematizations and governing techniques of urban risk reduction are embedded in, shaped by and contest narratives and processes that emerge from and circulate between different geographical scales. Scale helps to identify and explore how different scales of risk reduction are envisioned, produced, reproduced and altered between global and local contexts by international policy frameworks, non-governmental actors, authorities (of different governance levels) and citizens. Building on a case study of flood risks and risk governance in Mumbai, I argue that the specific way flood risks are governed in the city is the outcome of interlinked processes of problematizing flooding, as well as associated relations of power and knowledge, ranging from community and neighbourhood to global scales.

The next section delineates literature from different fields on the governance of disaster risks in, for and through the urban and then outlines the main conceptual underpinnings: geographical conceptualizations of scale and Foucault-inspired studies of regimes of practice. The fourth section justifies the case study selection and introduces the methods and data on which the chapter builds. Section five discusses the scalar politics of flood risk governance in Mumbai after 2005 through three empirically grounded examples: (1) decentralization as scalar fix; (2) scaled problematizations, rationalities and resulting governing techniques of flooding and (3) scalar interrelations created by and through acts, policies and different actor groups. Concluding thoughts on the multi-scalar configuration of disaster risk governance and geographical thinking are outlined in section six.

Studying the Governance of Disaster Risks in, for and through the Urban: Scale and Regimes of Practice

Research that studies how cities (can) produce, mitigate, navigate and cope with disaster risks has proliferated in recent years, drawing on different concepts and ranging from more normative to analytical in character. Scholars from various fields, such as urban political ecology, environmental and climate (in)justices and critical resilience studies, have investigated the creation and governance of risks at different scales. They highlight inequalities and power relations that can result in distinct levels of risk exposure (Parthasarathy 2009; Zeiderman 2012; Marks 2015), as well as unequal benefits and disadvantages arising from risk reduction measures or the trade-offs of respective policy agendas and actions (Chelleri et al. 2015; Alvarez and Cardenas 2019; Parthasarathy 2021; Ranganathan and Bratman 2021; Grove et al. 2024).

Research that more specifically assesses (climate-related) risk governance within and across scales in a globalized setting stems from the field of multilevel governance (Corfee-Morlot et al. 2009; Bulkeley 2010; Jänicke 2017: 114; GIZ 2018; Ishtiaque 2021). Such research engages with 'the connections between vertical tiers of government and horizontally organized forms of governance' (Betsill and Bulkeley 2006: 149) and has produced relevant insights into the scope, challenges and embeddedness of local governments' risk-related policies and measures. Yet, while the majority of research has focused on the processes and impacts of multilevel risk governance, the formation of these systems through social and political processes often remains underexamined (Wissen 2007: 233–4).

Other work has specifically examined the decentralization of fiscal, political and administrative powers, as responsibilities shift from national towards local governments (Rumbach 2016; Bhide 2017). Key disaster-related international policy documents support decentralization and emphasize the role of local governments in risk reduction and disaster management, suggesting that local governments should be empowered accordingly (UNISDR 2007, 2009, 2015). It has been found that decentralization can lead to more effective risk assessments and risk governance (Sharma et al. 2012; Rumbach 2016), but it can also hinder effective disaster risk governance in a city (Marks and Lebel 2016; Parthasarathy 2016). Since the 1990s, policymakers, practitioners and academics have encouraged further localization of risk reduction, particularly through community-based disaster risk management (Shaw 2012, 2016; Wolff 2021). However, critics argue that it is paradoxical to focus on reducing risks at local levels when these risks are usually caused by global or globally interconnected hazards and vulnerabilities (Clark-Ginsberg 2021).

Drawing on, building upon and extending the previously mentioned scholarly debates on urban risk governance, this chapter explores the production of urban risk governance within and across scales in Mumbai.

Scalar Perspectives on Urban Disaster Risk Governance

When speaking about or assessing scales in the context of risk governance, most prevailing literature considers scales as given and predefined (Aalders 2018: 94). That is, most scholarship seeks to identify appropriate scales for assessing or dealing with certain hazards such as municipal, regional, national or international levels. In contrast, human geographical thought emphasizes how scales are socially produced and therefore negotiable and interrelated. Using scale for the analysis of urban disaster risk governance, I undertake a deeper examination of how scales of risk and risk reduction are produced, negotiated, contested, and discursively and materially modified through social interactions among actors in relation to everyday practices, planning and decision-making (Swyngedouw 2004: 132; Ramasar 2014: 36).

Building on Sallie Marston's work on the production of social scale, four perspectives from scale theory are useful for this study. First, scales are socially constructed through political, economic, social and cultural processes within a particular social system rather than existing as an ontologically given category (Delaney and Leitner 1997; Marston 2000; Blackburn 2014). Second, considering scales as relational sheds light on the material and discursive connections between places and therefore on the spatial organization of society, both within a bounded geographical area and across geographical scales (Brenner 2001: 599–600). Third, although scalar configurations are (temporarily) 'fixed' and may persist for long periods of time, they were nonetheless produced by how society was understood at a particular moment in history (Marston 2004 in Blackburn 2014: 103). As such, long-term 'fixed' scalar configurations are also fluid and therefore changeable (Jonas 2006 in Blackburn 2014: 104).

Fourth, several intended and unintended processes and decisions concerning scales have been observed, which are considered under the politics of scale or scalar politics (MacKinnon 2010). Scale framing describes the 'discursive practices that construct meaningful (and actionable) linkages between the scale at which a social problem is experienced and the scale(s) at which it could be politically addressed or resolved' (Kurtz 2003: 894). Another relevant concept is scale jumping, which refers to processes whereby actors deploy strategies to bring concerns and political influence to other scales (MacKinnon 2010: 24). There is also scalar fixing, which depicts how certain scales become relevant temporarily or in the long term (Ramasar 2014: 39). Finally, processes through which the relevance of a scale is challenged, and new hierarchies of institutional organization, authority and regulation are generated have been termed rescaling (Brenner 2004: 449).

Scholars from human geography and related fields have employed constructivist, relational and political notions of scale in order to analyse urban risks. For example, previous studies have examined the role of scalar framing in mismatching individual risk exposure and scales of governmental actions in Nairobi (Aalders 2018), and how scalar framing helps to consider disjunctions between local political boundaries and suitable scales for urban disaster management in the Philippines (Gera 2018). With regard to the politics of scale in Jamaica's decentralized disaster risk governance, Sophie Blackburn (2014) argues that incomplete decentralization, scale jumping and scalar disconnect act as barriers to effective decentralization and to supporting the agency of local actors. Marks and Lebel (2016: 65) find that arguments about risks and responsibilities often invoke scale and level, both to promote and to resist disaster-related decentralizing efforts in Thailand. Considering scale-making as a 'relational product of how material, technical and discursive elements come to be assembled and reassembled in practice', Anders Blok (2016: 607) analyses how processes of urban change, including expert-led transformations towards climate resilience, are shaped but also contested through practices of knowledge-making in the Indian city of Surat.

Regimes of Practice

To understand scalar and scaled configurations in urban risk governance, this chapter also draws on Foucault's notion of governmentality (Foucault 2007). Governmentality is a specific 'art of governing', where government is understood as the 'conduct of conduct', either by others or through processes of self-governance (Dean 2009: 17). The forms of knowledge employed, and the dominant power relations that exist produce the specific art of governing and are therefore also important points of analysis. Governmentality-inspired studies do not focus on actors but rather consider the practices that constitute a government (Kammler et al. 2014). This notion of governmentality helps analyse how risks are problematized, how institutions, procedures and governing techniques enable risk reduction and, eventually, how urban disaster risk reduction becomes governable.

Governmentality may manifest as 'assemblage[s] of policies, strategies, and regulations that collectively define a dominant paradigmatic management approach' (Solecki et al. 2017: 1). An analysis of these assemblages, which Mitchell Dean (2009: 41) terms as 'regimes of practice', considers four key areas: the fields of visibility (Who and what is to be governed?); the technical aspects of government (How are authority and rule accomplished?); the episteme or rationality (How do thoughts and knowledge render problems governable?) and the formation of subjects and identities (What forms of person, self and identity are presupposed and elicited by practices of government?). Each of these components and the underlying power–knowledge–discourse structures are scaled and are produced, negotiated and altered through processes of scaling.

In recent years, particularly since 2010, numerous studies have emerged in the broad fields of climate governmentality (e.g. Oels 2005; Methmann 2010), resilience governmentality (Grove 2018b; O'Grady and Shaw 2023) and the government of risk, emergency and disasters (e.g. O'Malley 2009; Amoore 2013; Grove 2013; Bohle 2018; O'Grady 2018; Anderson 2021). These studies employ a Foucauldian lens to demonstrate and explain specific forms of governing that often align with shifts in rationalities. Some studies examine how the government of risks or disasters is linked to (neo)liberal arts of governing and biopolitics (e.g. Higgins 2001; Butler and Pidgeon 2011; Zeiderman 2013; Marchezini 2015; van Riet 2016; Grove 2018a; Lawrence 2018). Other studies focus on using governmentality to analyse how problematizations and rationalities are linked to particular techniques of governing (Escobar and Demeritt 2014; Ansell and Baur 2018; Bohle 2018; Becker 2021). For example, Nathaniel O'Grady (2018) depicts how emergency governance in the UK Fire and Rescue Service increasingly relies on ways of analysing, imagining and anticipating risks and future emergencies. Louise Amoore (2013: 1) argues that risk calculation is replaced by possibilistic risk modes in which uncertain futures are 'mapped and acted upon as possibilities' through technocratic expertise, thereby affecting processes of inclusion and exclusion. Some studies apply a governmentality lens to study climate change adaptation or disaster risk governance at urban or local levels (e.g. Rutland and Aylett 2008; Bulkeley and Schroeder 2012; Zeiderman 2012, 2013; Castán Broto 2017; Tozer 2018). Specifically relevant to this chapter are studies that explicitly approach the role of (elite and expert) knowledge in the (discursive) production of flood risk and how this knowledge is negotiated and contested through different arts of governing risks (Nygren 2016; Alvarez and Cardenas 2019; Ramalho 2019).

Towards a Conceptual Framework

In his work on governmentality, Foucault already considered spatial discourses and practices, such as the contextualization of problems or the selection of suitable

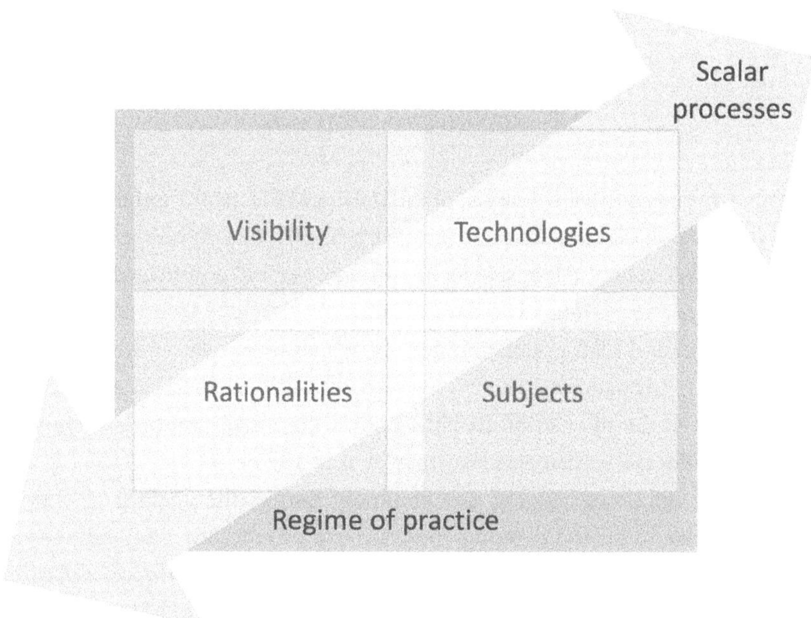

Figure 9.1 Illustration of conceptual framework.
Source: By author based on Dean's (2009) framework for the analysis of regimes of practices and human geographical notions of scale.

spaces of intervention, when analysing techniques of knowledge production and the exercise of power (Marquardt and Schreiber 2021: 196, 201). Yet I consider the combination of the two analytical angles useful to more specifically assess how disaster risk governance in, for and by the urban is produced in and across spaces.

Dean's analytical frame for the study of regimes of practices offers a comprehensive perspective on the constitution and quality of a disaster risk governance regime. In addition, notions of scale facilitate a more explicit study of how an 'urban' risk governance regime is a product of problematizations, rationalities, governing techniques and identities within and across distinct spatial scales, from self-governance to global governance, including the role of international institutions (Füller and Marquardt 2021: 95). As such, a combination of human geographic notions of scale and regimes of practice (see Figure 9.1) will be used to investigate (1) how the framing of flooding as a problem is linked to processes of scaling and rescaling; (2) how strategies, plans and governing techniques stem from and (temporarily) fix risks at different scales and (3) how actors produce, reproduce, alter or jump scales in order to make risks governable.

Case Site and Methods

Mumbai, the administrative capital of the state of Maharashtra and the financial capital of India, has a large commercial and trading base, establishing its presence in the global economy. Alongside its financial power and dominance, the city is also vulnerable to various everyday and disaster risks due to its coastal, low-lying location and high population density, which must rely on overburdened infrastructure (Appadurai 2001: 27; UN-HABITAT 2012; Sherly et al. 2015: 3). A large part of the population remains poor, with a high proportion living in hazardous informal settlements (Bardhan et al. 2015).

Flooding and waterlogging are common during the months of June–September due to low-lying topography and exposure to heavy rainfall during the monsoon season. Residents partly associate this period with the joy of playing in water and the opportunity to refill urban lakes as a source of water. However, for many residents, especially those living in marginalized communities and low-lying localities near bodies of water, flooding brings regular destruction, health burdens and other hardships such as power cuts, garbage accumulation and intermittent drinking water supply (Chatterjee 2010b; Bodke 2022). The tremendous rainfall on 25 and 26 July 2005 and the devastating floods that followed affected an unusually large proportion of Mumbai, including areas that are usually spared from flood hardships, although poor residents and those living in environmental risk-prone areas were affected at a higher rate (Parthasarathy 2009; Ravindran 2014). A total of 419 people were reported to have lost their lives in the flash floods through drowning, suffocation, electrocution, landslides and a stampede. An additional 216 persons died from flood-related illnesses (Gupta 2007: 189). Due to their scale and impacts on lives, livelihoods and the economy, the 2005 floods have been considered as a 'wake-up call' and a catalyst for change in disaster risk governance (Zimmermann 2019). As the frequency and intensity of extreme rainfall events are likely to increase in the future, flood risks will persist and even possibly increase in intensity, leaving 35 per cent of Mumbai's population exposed to flood risk – in some wards as much as 60 per cent (Ranger et al. 2011; IPCC 2021; BMC 2022a,b). Given its prominent status as India's economic centre and its exposure to a variety of risks exacerbated by climate change, Mumbai has received increasing attention from international media (Ravindran 2014; Gettleman 2017; Kumar 2019), scientists (e.g. Petzold et al. 2024) and circles of experts and advisers on urban resilience (C40 Cities 2022; Climate Champions and Fazeli 2023), all of which bring distinct perspectives and interests. In this context, Mumbai presents an interesting case for exploring the multi-scalar configuration of flood risk governance.

I conducted three study trips to Mumbai: from October 2015 to January 2016, from January 2019 to May 2019, and in November 2022. A longitudinal approach to the research allows for the observance of changes in discourse, strategies, measures and inter-scalar relations over time. Semi-structured and problem-centred interviews were conducted with officials and bureaucrats at ward, district, city, state and federal levels. I also spoke with academic researchers, activists, journalists, representatives of non-governmental organizations (NGOs) and members of community-based organizations (CBOs). Field visits were mainly carried out along the Dahisar river in the north-western suburbs of Borivali and Dahisar. The field visits comprised transect walks, photographic documentation and interviews and ad hoc group discussions with residents of flood-prone neighbourhoods. Discussions focused on the perception, implementation and negotiation of flood governance techniques at ward and neighbourhood levels. A systematic analysis of policies, plans and guidelines from the fields of disaster (risk) management, climate change adaptation and development planning at various levels was conducted. In addition, newspaper articles and notes from thematically relevant events such as conferences, workshops or seminars were included in the analysis.

Governing Floods in Mumbai – Scale, Problematizations and Governing Techniques

Building on the conceptual framework linking scale and regimes of practice, the following three sections address the scalar politics of flood risk governance in Mumbai after 2005 through three analytical angles: (1) decentralization of disaster and risk governance as a scalar fix; (2) the way flooding is rationalized and scaled as a problem and (3) scalar interrelations created by and through policies and different actor groups.

Decentralized Governance Structures and Implications for Making Flood Risks Governable

This section traces how decentralized approaches to flood risk governance have been encouraged through laws, policies and plans in the fields of disaster risk reduction and climate change adaptation. The results of the decentralization processes, in the form of strengthened risk governance responsibilities at state, district and urban level, can be considered a 'scalar fix', that is, a relatively stable (temporary) spatial and scalar form of organization (Herod 2011: 28–9 in Ramasar 2014: 39).

In India, reforms in the context of neoliberal economic and political restructuring aimed at improving economic performance have redistributed powers between national state governments, municipal bodies and municipal ward-level committees since the 1990s (Jain et al. 2021). In the case of disaster governance, national ministries were long responsible for handling disasters (Chatterjee 2010a; Gupta 2018) before the Disaster Management Act (DMA 2005) provided an institutional mechanism and mandate to decentralize disaster (risk) management to the State and District Disaster Management Authorities (Government of India 2005) (see Figure 9.2). As such, at present, the Maharashtra State Disaster Management Unit publishes State Disaster Management Plans and guides the districts in matters of risk mitigation and preparedness (GoM 2023). In Mumbai, the Disaster Management Department (DMD) of the Municipal Corporation of Greater Mumbai (BMC) directs disaster risk-related

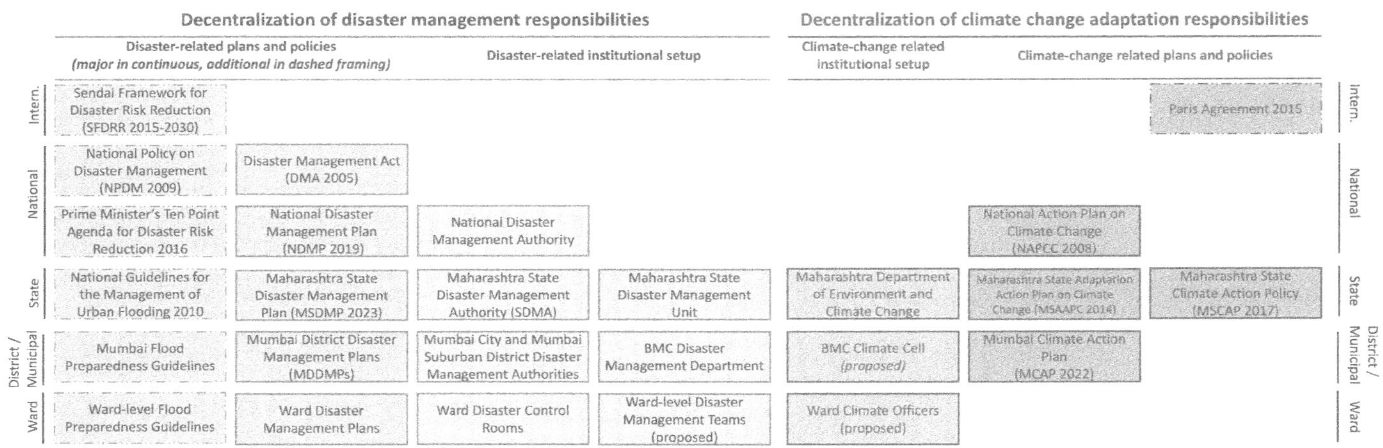

Figure 9.2 Organigram of decentralized responsibilities for flood risk governance in Mumbai.
Source: Author.

activities jointly for the Mumbai Suburban and Mumbai City District Disaster Management Authorities. The major floods in 2005 further acted as a catalyst for decentralization.[2] The Greater Mumbai Disaster Management Action Plan, released in 2007, advised the twenty-four wards to develop ward-level disaster management plans (Government of Maharashtra 2007). The central control room, operated by the DMD, was upgraded, and employees could now monitor CCTV cameras and automatic weather stations, disseminate warnings, receive emergency calls and coordinate emergency measures between the twenty-four ward control rooms and emergency operators around the clock. In recent years, the Disaster Management Plans for Mumbai Suburban and Mumbai City districts have been regularly updated (see, e.g. DDMA 2019). In 2022, the corporation announced its plan to establish decentralized disaster management cells in each ward through the deployment of ward-level disaster management teams (Express News Service 2022). These twelve-member teams are expected to analyse data, identify potential risks, prepare action plans and sensitize citizens about disaster risks (FPJ News Service 2024). In climate change adaptation (an area also relevant to flood risk governance), it has also been encouraged to develop plans and measures at state levels and below, resulting in Maharashtra's State Adaptation Action Plan on Climate Change (MSAAPC) in 2014 and the release of Mumbai's first Climate Action Plan (MCAP) in 2022 (GoI 2008; TERI 2014; BMC 2022b). The MCAP proposed the establishment of climate cells within the BMC Department of Environment, which is to be renamed the Department of Environment and Climate Change, and the creation of posts for climate officers in all twenty-four wards (BMC 2022b: 190–1).

This decentralization and localization of governance structures is also supported by international policy frameworks. The SFDRR specifically calls for an increase of the 'number of countries with national and local disaster risk reduction strategies' (UNISDR 2015, Target E). The NUA further specifies the relevance of reducing disaster risks and enhancing resilience, particularly for urban areas (see NUA §13 and 14). The SDGs call to 'develop and implement . . . holistic disaster risk management at all levels', specifically for cities and human settlements (SDGs 11.5 and 11.b). The Paris Climate Agreement emphasizes the relevance of local-level engagement in climate change mitigation and adaptation. The shifting of powers and responsibilities to subnational and local governments can be an effective means to enhancing disaster response and risk governance mechanisms. Yet the existence of 'multiple planning agencies for different functions, at different levels of aggregation and scale, which prevent integration of objectives and strategies'

(Parthasarathy 2016: 28; see also map by Kamath and Tiwari 2022: 679), together with the lack of a decentralized institutional framework that fosters interconnection between disaster risk reduction and climate change adaptation across scales (Weinstein et al. 2019; Zimmermann et al. 2023), can also pose challenges for integrated flood risk mitigation. The outcomes of decentralization processes, and the way in which international and national networks, private businesses and civil society groups use the resulting spaces to share their expertise and ideas while also promoting their own interests in the process, and the implications for multi-scalar decision-making processes require further assessment.

Scaling the Problem of Flooding – Rationalities, Framings and Problematization

Building on a Foucauldian understanding of regimes of practices (Dean 2009), this section outlines how the problem of flooding is visualized, rationalized and thereby framed at different scales and how this plays out in governing techniques in Mumbai. As systems of consensus of thought, rational frames of references are so well established that they are considered natural and objective (Mattissek 2010: 132). A way to approach rationalities is through assessing problematization, that is, the processes and forms of knowledge and thought through which a (complex) phenomenon is framed as a problem and turned into an object of government (Foucault 1998; Dean 2009; Hutter et al. 2014: 275; Wakefield 2020; Collier and Gruendel 2022: 3).

The following paragraphs focus on two seemingly contrasting ways of framing the problem of flooding in Mumbai that exemplify the spatiality and scalar relations of governing flood risks. First, to a certain extent, a framing of the phenomenon of flooding has emerged that scales the problem to specific localities and times of the year, thereby disentangling it from more complex questions of urban development and social vulnerability. In recent years, flooding in Mumbai has been experienced and portrayed as a seasonal problem and a localized phenomenon that occurs along water bodies and in low-lying flood-prone areas. Although city-wide governing techniques such as enhanced Standard Operating Procedures and rainfall forecasting and warning have received increasing attention since 2005 (Zimmermann et al. 2023), there also exist several localized approaches, including the installation of pumping stations, the construction of retaining walls or the demarcation of flood-prone spots.

These governing techniques are backed by a rationale that attributes the causes of flooding (apart from the extraordinary rainfall amounts) to the city's topography, insufficient and blocked stormwater drainage, and encroachment on riverbanks. The flooding spots, which have been demarcated on maps and are defined within the annually published Flood Preparedness Guidelines (BMC 2022b), are highly relevant and can affect a locality's access to attention, finance and technologies. For example, the number of tackled versus remaining flooding spots is used to underline the administration's achievements with regard to flood risk governance (Pinto 2020; Parab-Pandit 2022; Surve 2022). Furthermore, selected 'flooding hotspots' have been chosen for the installation of underground water storage tanks, which were built based on flood tunnels in Tokyo, Japan and underground ponds in South Korea (Singh 2020, 2021; Parab-Pandit 2023). Experts have expressed scepticism about the hotspot strategy, as it could result in shifting flooding and waterlogging risks to nearby areas. Instead, they argue that more data are needed to understand and approach flooding at other scales (Gokarn 2024).

Mumbai's Development Plan 2014–34 proposes the demarcation of buffer spaces as development-free zones along rivers, creeks and canals but also allows for the opening of previously designated No Development Zones for construction (Mulye 2018; PNT Bureau 2024). Furthermore, activists and researchers have pointed out that the planning processes of several large-scale infrastructure projects have not fully considered their impacts on flooding (STUP Consultants and Ernst & Young 2016; Anand 2019; Anthony 2022). Therefore, while the framing of flooding as a local problem for certain neighbourhoods and localities can lead to effective locally based solutions, this understanding may also promote a detachment from more complex questions of urban development and holistic planning approaches.

A second problematization of flood risks has gained prominence in recent years, which frames the 'problem' of flooding in the context of climate change. While this problematization is partly linked to the first – for example, through studies suggesting that changing rainfall patterns will contribute to more localized flooding and also flooding outside the monsoon season (BMC 2022a) – the increasing focus on climate change as a contributor to urban flooding also stands apart from it, as it locates the origins of the problem (or at least part of it) at a different, more distant scale. Although flooding is not novel in Mumbai, studies which assess flood risks in the context of climate change have flourished in recent years (Ranger et al. 2011; Mehta et al. 2019; Kumar et al.

2020; Murali et al. 2020). While Mumbai has held a prominent position in both academic and policy-oriented debates on urbanization and the vulnerability of coastal (mega)cities for many years (UN-HABITAT 2012; GreenAsh 2013; Peters et al. 2015; Sherly et al. 2015; Chouhan et al. 2017), it can be assumed that, through the framing of climate change, flood-related governing techniques will require even greater justification for the international community of experts and policymakers observing the management of risks in Mumbai. In 2020, Mumbai became a member of the C40 network and released its first Climate Action Plan in 2022, which was developed by the Brihanmumbai Municipal Corporation (BMC) with technical support from the World Resources Institute (WRI) India (BMC 2022b). The plan explicitly describes the outcomes of climatic changes for Mumbai and defines steps to alleviate the worst impacts. As such, it proves to the international circle of experts that climate change is being taken seriously. The cities' wish to be recognized as a forerunner in climate action is explicit in this document, as one of the Leaders' Pledges at the beginning of the document emphasizes: 'With the launch of the Mumbai Climate Action Plan, Mumbai has set the benchmark for Climate Action in India' (Shri. Aaditya Thackeray in BMC 2022b: 9). With increasing attention on urban risks and resilience, being regarded as a pioneer or role model can become an important qualifying characteristic for cities worldwide (Wakefield 2020: 3; Kamath and Tiwari 2022). Meanwhile, the increasing framing of flooding in the context of climate change, along with the financial and knowledge-based resources attached to it, has also been used to link flood risk governance to questions of climate change adaptation and development planning. For example, the City Disaster Management Institute has developed and tested training modules on climate change (GoM 2023: 191); the BMC has newly established an Environment and Climate Change Department with urban flooding as one of its priority areas (BMC 2024b) and has initiated climate budgeting (BMC 2024a).

Scalar Interrelations

As the two previously discussed tendencies of problematizing flood risks suggest, power–knowledge relations at and among different scales shape how flood risks are framed and how flooding is governed in the city. This section further depicts scalar interrelations by giving examples of how international attention and policy frameworks enable local actors and communities to link their claims to globally circulating knowledge. Aligned processes of scale jumping and scalar configuration can be considered as mechanisms of 'political strategy' (Smith 1993; Brown and Purcell 2005) or 'bottom-up resistance' (Blackburn 2014). That is, they are mechanisms that aim to challenge or evade existing scalar structures in the hope of creating more favourable conditions for pursuing intended goals. In regard to risk reduction and resilience building, scalar interrelations have been observed and discussed in various cases across the world, for example, in the synchronization of 'top-down' government action or knowledges and 'bottom-up' improvisation and community involvement (DeVerteuil et al. 2021; O'Grady et al. 2022; O'Grady and Shaw 2023), as well as in attempts to enable critique and enhance aspects of agency and justice (Cox et al. 2022; Grove et al. 2024).

When examining the case of Mumbai, two examples serve to illustrate scalar interrelations. Local-level state actors both access international (climate) expertise and use international policy documents, such as the Sendai Framework, as reference points, connecting local needs or ideas to global discourse and strategies. As a disaster management official from another district argued, referring to the Sendai Framework can help to convince supervisors and politicians of new ideas and measures:

> While implementing a new thing, we need references. Means, I am implementing at district level and you are my boss. You will ask that is there any guideline for this? Then immediately I can use the document for that, that it is mentioned for the Sendai Framework.[3]

In this context, participation in (international) workshops and conferences and connection with academia can be a strategy used by local officials to access knowledge and resources, and to feel empowered, give meaning and provide justifications for intended risk-governing techniques. Other prominent examples of scalar interrelations and scale jumping used as political strategy include Mumbai's membership in C40 since 2020, the global network 'of the world's leading cities' confronting the climate crisis (C40 Cities 2024), and the BMC's collaboration with the WRI, a globally well-connected think-tank that provided technical support for MCAP.

International attention and scalar interrelations also give options for risk-prone communities to rescale their problems through transdisciplinary and international alliances. For example, residents and CBOs in a north-western suburb approached actors at other levels with their flood-related claims. As a result, they rescaled the problem of flooding from the neighbourhood scale to a wider relevance for actors at higher scales. The River March Group, a public movement aiming at restoring and rejuvenating Mumbai's environment, connected with the Maharashtra State government in order to get permission for the construction of check dams to increase water percolation and prevent flooding on the Dahisar and Poisar rivers and approached the Art of Living Foundation to acquire the funds needed (Mumbai March 2020). Residents of a housing society in the same ward approached the central government to get permission to change development controls in the area in an effort to enable the construction of more flood-proof apartments.[4] Another example of scale jumping is the Koli fishing communities in Mumbai, who participated in a research project centred on examining marginal environments characterized by high levels of uncertainty (Tapestry Project 2024). Through their participation, their perspectives and concerns regarding climate change impacts, threats from human disturbance and adaptation options became prominently featured in the media (The Quint 2022) and academic publications (Mehta et al. 2019; Adam et al. 2022).

Conclusion and Outlook

Building on the constructivist notion of scale and on empirical insights from Mumbai, I suggest that the specific way flood risks are governed in the city is the outcome of processes of problematizing flooding, as well as associated relations of power and knowledge, which are spatially interlinked across scales. The examination of processes of decentralization, forms of problematizing and governing flooding, and the scalar interrelations created by and through acts, policies and different actor groups shows that diverse forms of knowledge, expertise and ways of framing risks interrelate in Mumbai's flood risk governance regime, spanning from community and neighbourhood to global scales. Therefore, I suggest that it is not sufficient to understand urban disaster risk governance as the sum of governing arrangements present in a city, but that it can rather be conceptualized as a multi-scalar configuration of power and knowledge.

The empirical insights provided above, though far from giving a complete picture, exemplify how this configuration is embedded in and shaped by processes occurring in, at and between various scales. Based on the insights into decentralization tendencies, problem framing and scalar interrelations, two observations can be made. First, local knowledge as well as scientific and technocratic expertise – which are embedded in internationally circulating discourses on risks and disasters – are employed in attempts to find solutions to specific challenges arising in certain localities. The resulting implications on power relations and processes of subjectivation need further assessment. While it may become easier for residents of flood-prone neighbourhoods to make their claims with regard to their flood exposure, it can become more difficult to inquire into, challenge or contest measures developed or decisions taken that build on advanced scientific knowledge or international expertise. Second, flood risks and their reduction are embedded in global relations of both physical flows (in terms of ongoing climate change and respective changes in precipitation patterns and sea levels) and knowledge and power (in terms of framing and identifying suitable ways of understanding and responding to risks). With the ongoing implementation of international policy frameworks and the associated forms of knowledge and expertise, it is likely that the global circulation of ideas, governing techniques and subjects will increasingly contribute to shaping urban processes of risk reduction and resilience building in Mumbai and other cities. The proposed conceptual framework opens an analytical perspective for assessing the resulting regimes of practice in their configurations across geographical scales.

Based on the illustration of the potentials of understanding regimes of urban risk reduction as scalar configurations, I argue that this combined perspective opens space to (1) identify points of intersection between different scales of problematizing and governing flood risks; (2) understand risk management regimes as embedded in and product of forms of knowledge, rationalities and governing techniques across various scales; (3) examine how global policy frameworks and discourses unfold and are utilized in local realities and (4) identify potentials for transformation, for example, by analysing how new alliances across scales influence power–knowledge relations and vice versa. The findings of this chapter suggest that human geographical research focusing on risk creation and risk reduction and on role of power–knowledge relations in shaping these processes can contribute to a better understanding of the intersection of urbanization, disasters and risk governance.

Notes

1 For example, UN Making Cities Resilient Campaign; ICLEI Resilient Cities; Rockefeller Foundation 100 Resilient Cities; C40; Asian Cities Climate Change Resilience Network – ACCCRN.

2 Interview with a city official in 2015.

3 Interview with the district disaster management officer in 2019.
4 Interview with a resident of the R-central ward in 2019.

References

Aalders, J. T. (2018), 'The Scale of Risk: Conceptualising and Analysing the Politics of Sacrifice Scales in the Case of Informal Settlements at Urban Rivers in Nairobi', *Erdkunde* 72 (2): 91–101. https://doi.org/10.3112/erdkunde.2018.02.02.

Adam, H. N., S. Movik, D. Parthasarathy, A. Alankar, N. C. Narayanan and L. Mehta (2022), 'Climate Change and Uncertainty in India's Maximum City, Mumbai', in L. Mehta, H. N. Adam, S. Srivastava and L. Mehta (eds), *The Politics of Climate Change and Uncertainty in India: Abingdon Oxon* (Pathways to Sustainability), 134–60, Routledge.

Alvarez, M. K. and K. Cardenas (2019), 'Evicting Slums, "Building Back Better": Resiliency Revanchism and Disaster Risk Management in Manila', *International Journal of Urban and Regional Research* 43 (2): 227–49. https://doi.org/10.1111/1468-2427.12757.

Amoore, L. (2013), *The Politics of Possibility: Risk and Security Beyond Probability*, Duke University Press.

Anand, N. (2019), 'Mumbai's Coastal Reclamation Project – The Likelihood of a Not-So-Natural Disaster', *Wire*, 11 April. Available at: https://science.thewire.in/society/urban/mumbais-coastal-reclamation-project-the-likelihood-of-a-not-so-natural-disaster/ (accessed 4 September 2024).

Anderson, B. (2021), 'Scenes of Emergency: Dis/Re-assembling the Promise of the UK Emergency State', *Environment and Planning C: Politics and Space* 39 (7): 1356–74. https://doi.org/10.1177/2399654420954214.

Anguelovski, I., L. Shi, E. Chu, D. Gallagher, K. Goh, Z. Lamb, K. Reeve and H. Teicher (2016), 'Equity Impacts of Urban Land Use Planning for Climate Adaptation: Critical Perspectives from the Global North and South', *Journal of Planning Education and Research* 36 (3): 333–48.

Ansell, C. and P. Baur (2018), 'Explaining Trends in Risk Governance: How Problem Definitions Underpin Risk Regimes', *Risk, Hazards & Crisis in Public Policy* 9 (4): 397–430. https://doi.org/10.1002/rhc3.12153.

Anthony, H. (2022), 'Despite Promises by Political Parties, Annual Mumbai Floods Still a Problem', *Citizen Matters*, 21 January. Available at: https://citizenmatters.in/despite-promises-by-political-parties-annual-mumbai-floods-still-a-problem/ (accessed 4 December 2024).

Appadurai, A. (2001), 'Deep Democracy: Urban Governmentality and the Horizon of Politics', *Environment & Urbanization* 13 (2): 23–44.

Bardhan, R., S. Sarkar, A. Jana and N. R. Velaga (2015), 'Mumbai Slums Since Independence: Evaluating the Policy Outcomes', *Habitat International* 50:1–11. https://doi.org/10.1016/j.habitatint.2015.07.009.

Becker, P. (2021), 'Fragmentation, Commodification and Responsibilisation in the Governing of Flood Risk Mitigation in Sweden', *Environment and Planning C: Politics and Space* 39 (2): 393–413. https://doi.org/10.1177/2399654420940727.

Betsill, M. M. and H. Bulkeley (2006), 'Cities and the Multilevel Governance of Global Climate Change', *Global Governance* 12:141–59.

Bhide, A. (2017), 'Directed Decentralisation: Analysing the Experience of Decentralisation via JNNURM in Maharashtra', in N. Jayaram (ed.), *Social Dynamics of the Urban: Studies from India*, (Exploring Urban Change in South Asia), Bd. 35, 81–97, Springer.

Blackburn, S. (2014), 'The Politics of Scale and Disaster Risk Governance: Barriers to Decentralisation in Portland, Jamaica', *Geoforum* 52:101–12. https://doi.org/10.1016/j.geoforum.2013.12.013.

Blok, A. (2016), 'Assembling Urban Riskscapes', *City* 20 (4): 602–18. https://doi.org/10.1080/13604813.2016.1194000.

BMC (2022a), 'Climate & Air Pollution Risks and Vulnerability Assessment for Mumbai, India: March 2022', Study Conducted by the World Resources India, Ross Centre for Sustainable Cities, in Collaboration with the Brihanmumbai Municipal Corporation.

BMC (2022b), 'Mumbai Climate Action Plan 2022: Brihanmumbai Municipal Corporation. Mumbai'. Available at: https://mcap.mcgm.gov.in/about/ (accessed 2 September 2021).

BMC (2024a), 'Climate Budget Report: FY 2024–25', Hg. v. Brihanmumbai Municipal Corporation. Available at: https://mcap.mcgm.gov.in/wp-content/uploads/2024/06/English_Climate-Budget-Report-2024-25_compressed.pdf (accessed 4 September 2024).

BMC (2024b), 'Environment and Climate Change Department'. Available at: https://portal.mcgm.gov.in/irj/portal/anonymous/qlenvironment (accessed 4 September 2024).

Bodke, S. (2022), 'Coping with Monsoons in an Informal Settlement', *Citizen Matters*, 13 October. Available at: https://citizenmatters.in/kranti-nagar-informal-settlement-monsoons/ (accessed 4 September 2024).

Bohle, J. (2018), 'Hurricane-Riskscapes and Governmentality', *Erdkunde* 72 (2): 125–34. https://doi.org/10.3112/erdkunde.2018.02.04.

Brenner, N. (2001), 'The Limits to Scale? Methodological Reflections on Scalar Structuration', *Progress in Human Geography* 25 (4): 591–614. https://doi.org/10.1191/030913201682688959.

Brenner, N. (2004), 'Urban Governance and the Production of New State Spaces in Western Europe, 1960–2000', *Review of International Political Economy* 11 (3): 447–88. https://doi.org/10.1080/0969229042000282864.

Brown, J. C. and M. Purcell (2005), 'There's Nothing Inherent About Scale: Political Ecology, the Local Trap, and the Politics of Development in the Brazilian Amazon', *Geoforum* 36 (5): 607–24. https://doi.org/10.1016/j.geoforum.2004.09.001.

Bulkeley, H. (2010), 'Cities and the Governing of Climate Change', *Annual Review of Environment and Resources* 35 (1): 229–53. https://doi.org/10.1146/annurev-environ-072809-101747.

Bulkeley, H. and H. Schroeder (2012), 'Beyond State/Non-State Divides: Global Cities and the Governing of Climate Change', *European Journal of International Relations* 18 (4): 743–66. https://doi.org/10.1177/1354066111413308.

Butler, C. and N. Pidgeon (2011), 'From "Flood Defence" to "Flood Risk Management": Exploring Governance, Responsibility, and Blame', *Environment and Planning C: Politics and Space* 29 (3): 533–47. https://doi.org/10.1068/c09181j.

C40 Cities (2022), 'Mumbai Climate Action Plan: A Deep Dive into the First Climate Action Plan from a C40 City in the South and West Asia Region'. Available at: https://www.c40.org/news/mumbai-climate-action-plan/ (accessed 6 November 2024).

C40 Cities (2024), 'About C40'. Available at: https://www.c40.org/about-c40/ (accessed 9 September 2024).

Castán Broto, V. (2017), 'Urban Governance and the Politics of Climate Change', *World Development* 93:1–15. https://doi.org/10.1016/j.worlddev.2016.12.031.

Chatterjee, M. (2010a), 'Resilient Flood Loss Response Systems for Vulnerable Populations in Mumbai: A Neglected Alternative', PhD diss., Rutgers University, New Brunswick.

Chatterjee, M. (2010b), 'Slum Dwellers Response to Flooding Events in the Megacities of India', *Mitigation and Adaptation Strategies for Global Change* 15 (4): 337–53. https://doi.org/10.1007/s11027-010-9221-6.

Chelleri, L., J. J. Waters, M. Olazabal and G. Minucci (2015), 'Resilience Trade-Offs: Addressing Multiple Scales and Temporal Aspects of Urban Resilience', *Environment & Urbanization* 27 (1): 181–98. Available at: https://journals.sagepub.com/doi/full/10.1177/0956247814550780 (accessed 6 November 2024).

Chouhan, H. A., D. Parthasarathy and S. Pattanaik (2017), 'Urban Development, Environmental Vulnerability and CRZ Violations in India: Impacts on Fishing Communities and Sustainability Implications in Mumbai Coast', *Environment, Development and Sustainability* 19 (3): 971–85. https://doi.org/10.1007/s10668-016-9779-6.

Chu, E. K. (2018), 'Transnational Support for Urban Climate Adaptation: Emerging Forms of Agency and Dependency', *Global Environmental Politics* 18 (3): 25–46. https://doi.org/10.1162/glep_a_00467.

Clark-Ginsberg, A. (2021), 'Risk Technopolitics in Freetown Slums', in J. A. C. Remes and A. Horowitz (eds), *Critical Disaster Studies* (Critical Studies in Risk and Disaster), 85–96, University of Pennsylvania Press.

Climate Champions and H. Fazeli (2023), 'From Mumbai to Paris: How Cities Are Spearheading Adaptation Action', Hg. v. UNFCCC Climate Champions. Available at: https://climatechampions.unfccc.int/from-mumbai-to-paris-how-cities-are-spearheading-adaptation-action/ (accessed 6 November 2024).

Collier, S. J. and A. Gruendel (2022), 'Design in Government: City Planning, Space-Making, and Urban Politics', *Political Geography* 97:102644. https://doi.org/10.1016/j.polgeo.2022.102644.

Corfee-Morlot, J., L. Kamal-Chaoui, M. G. Donovan, I. Chochran, A. Robert and P.-J. Teasdale (2009), 'Cities, Climate Change and Multilevel Governance', Hg. v. OECD (OECD Environment Working Papers, 14). https://doi.org/10.1787/220062444715.

Cox, S., K. Grove and A. Barnett (2022), 'Design-Driven Resilience and the Limits of Geographic Critique', *Geographical Journal* 188 (2): 294–308. https://doi.org/10.1111/geoj.12437.

DDMA (2019), 'District Disaster Management Plan', Mumbai City District Disaster Management Authority, Mumbai City District.

Dean, M. (2009), *Governmentality: Power and Rule in Modern Society*, 2nd edn, reprint, Sage.

Delaney, D. and H. Leitner (1997), 'The Political Construction of Scale', *Political Geography* 16 (2): 93–7.

DeVerteuil, G., O. Golubchikov and Z. Sheridan (2021), 'Disaster and the Lived Politics of the Resilient City', *Geoforum* 125:78–86. https://doi.org/10.1016/j.geoforum.2021.07.004.

Disaster Management Act (2005), 'Government of India'. Available at: https://ndmindia.mha.gov.in/ndmi/images/The%20Disaster%20Management%20Act,%202005.pdf (accessed 28 October 2025).

Escobar, M. P. and D. Demeritt (2014), 'Flooding and the Framing of Risk in British Broadsheets, 1985–2010', *Public Understanding of Science (Bristol, England)* 23 (4): 454–71. https://doi.org/10.1177/0963662512457613.

Express News Service (2022), 'BMC Considering Proposal to Set Up Ward-Level Disaster Management Cells', *Indian Express*, 15 March. Available at: https://indianexpress.com/article/cities/mumbai/bmc-considering-proposal-to-set-up-ward-level-disaster-management-cells-7821424/ (accessed 4 September 2024).

Foucault, M. (1998), 'Polemics, Politics and Problematizations: Interview by P. Rabinow, May 1984', in P. Rabinow (ed.), *Essential Works of Foucault*, vol. 1, 111–19, New Press. Available at: https://foucault.info/documents/foucault.interview/ (accessed 22 November 2024).

Foucault, M. (2007), *Security, Territory, Population: Lectures at the College De France, 1977–1978*, Palgrave Macmillan.

FPJ News Service (2024), 'Mumbai: BMC to Appoint Contractual Staff for Ward-Level Disaster Management Amid Recruitment Delays', *Free Press Journal*, 16 July. Available at: https://www.freepressjournal.in/mumbai/mumbai-bmc-to-appoint-contractual-staff-for-ward-level-disaster-management-amid-recruitment-delays (accessed 4 September 2024).

Füller, H. and N. Marquardt (2021), '3 Gouvernementalität in der humangeographischen Diskursforschung', in G. Glasze (ed.), *Handbuch Diskurs und Raum. Theorien und Methoden für die Humangeographie sowie die sozial- und kulturwissenschaftliche Raumforschung. Unter Mitarbeit von Annika Mattissek* (Sozial- und Kulturgeographie), 3rd edn, vol. 11, 87–108, transcript verlag.

Fünfgeld, H. (2015), 'Facilitating Local Climate Change Adaptation Through Transnational Municipal Networks', *Current Opinion in Environmental Sustainability* 12:67–73. https://doi.org/10.1016/j.cosust.2014.10.011.

Fünfgeld, H. and B. Schmid (2020), 'Justice in Climate Change Adaptation Planning: Conceptual Perspectives on Emergent Praxis', *Geographica Helvetica* 75:437–49.

Gera, W. (2018), 'Scalar Politics in Philippine Urban Disaster Management: Reframing Metropolitan Governance for Local Resilience and Sustainability', *Erdkunde* 72 (4): 287–311. https://doi.org/10.3112/erdkunde.2018.04.03.

Gettleman, J. (2017), 'Coping with Floods Is Only Half the Battle for Mumbai's Poor', *New York Times*, 30 August. Available at: https://www.nytimes.com/2017/08/30/world/asia/mumbai-monsoon-flood-houston.html (accessed 6 November 2024).

GIZ (2018), 'Multi-Level Climate Governance Supporting Local Action: Instruments Enhancing Climate Change Mitigation and Adaptation at the Local Level', Hg. v. Deutsche Gesellschaft Für Internationale Zusammenarbeit (GIZ) GmbH, Bonn.

GoI (2008), 'National Action Plan on Climate Change: Prime Minister's Council on Climate Change', New Delhi. Available at: https://moef.gov.in/uploads/2018/04/NAP_E.pdf (accessed 12 July 2025).

Gokarn, S. (2024), 'Mumbai Monsoon: Understanding BMC's Flood Mitigation Measures and What to Expect', *Citizen Matters*, 2 July. Available at: https://citizenmatters.in/mumbai-flood-mitigation-bmc/ (accessed 4 September 2024).

GoM (2023), 'State Disaster Management Plan 2023', Disaster Management, Relief and Rehabilitation Department, Government of Maharashtra.

Government of India (2005), 'The Disaster Management Act, 2005', The Gazette of India Part II, Section 1. Available at: https://moef.gov.in/uploads/2018/04/NAP_E.pdf (accessed 28 October 2025).

Government of Maharashtra (2007), *Greater Mumbai Disaster Management Action Plan: Risk Assessment and Response Plan*, vol. 1. Available at: https://karmayog.org/wp-content/uploads/10392/15917.pdf (accessed 28 October 2025).

GreenAsh (2013), 'Natural Disaster Risk Levels of the World's Largest Cities'. Available at: http://greenash.net.au/thoughts/2013/03/natural-disaster-risk-levels-of-the-worlds-largest-cities/ (accessed 20 June 2020).

Grove, K. (2013), 'From Emergency Management to Managing Emergence: A Genealogy of Disaster Management in Jamaica', *Annals of the Association of American Geographers* 103 (3): 570–88.

Grove, K. (2018a), 'Disaster Biopolitics and the Crisis Economy', in J. L. Lawrence and S. M. Wiebe (eds), *Biopolitical Disaster* (Interventions), 30–46, Routledge/Taylor & Francis Group.

Grove, K. (2018b), *Resilience* (Key Ideas in Geography), Taylor & Francis.

Grove, K., G. Reid, S. Molinari, J. Falcon, A. Mehta-Kroll, E. Sosa El Fakih, A. Sepulveda-Reyes and D. Ortiz (2024), 'Absurd Geographies of Resilience and Justice', *Climate and Development* 16 (9): 739–50. https://doi.org/10.1080/17565529.2023.2255566.

Gupta, A. K. (2018), 'Disaster Governance and Legal Systems in India', in I. Pal and R. Shaw (eds), *Disaster Risk Governance in India and Cross Cutting Issues* (Disaster Risk Reduction, Methods, Approaches and Practices), 39–60, Springer Singapore.

Gupta, K. (2007), 'Urban Flood Resilience Planning and Management and Lessons for the Future: A Case Study of Mumbai, India', *Urban Water Journal* 4 (3): 183–94. https://doi.org/10.1080/15730620701464141.

Higgins, V. (2001), 'Calculating Climate: "Advanced Liberalism" and the Governing of Risk in Australian Drought Policy', *Journal of Sociology* 37 (3): 299–316.

Hutter, G., M. Leibenath and A. Mattissek (2014), 'Governing Through Resilience? Exploring Flood Protection in Dresden, Germany', *Social Sciences* 3 (2): 272–87. https://doi.org/10.3390/socsci3020272.

IPCC (2021), 'Summary for Policymakers', in V. Masson-Delmotte, P. Zhai, A. Pirani, S. L. Connors, C. Péan, S. Berger, et al. (eds), *Climate Change 2021: The Physical Science Basis: Contribution of Working Group I to the Sixth Assessment Report of the Intergovernmental Panel on Climate Change*, 3–32, Cambridge University Press.

Ishtiaque, A. (2021), 'Multilevel Governance in Climate Change Adaptation: Conceptual Clarification and Future Outlook', in A. Fares (ed.), *Climate Change and Extreme Events*, 171–85, Elsevier.

Jain, M., A. Korzhenevych and R. Hecht (2021), 'Two Decades of Urban and Rural Restructuring in India: An Empirical Investigation Along Delhi-Mumbai Industrial Corridor', *Habitat International* 117:102444. https://doi.org/10.1016/j.habitatint.2021.102444.

Jänicke, M. (2017), 'The Multi-level System of Global Climate Governance – The Model and Its Current State', *Environmental Policy and Governance* 27 (2): 108–21. https://doi.org/10.1002/eet.1747.

Kamath, L. and A. Tiwari (2022), 'Ambivalent Governance and Slow Violence in Mumbai's Mithi River', *International Journal of Urban and Regional Research* 46 (4): 674–86. https://doi.org/10.1111/1468-2427.13092.

Kammler, C., R. Parr and U. J. Schneider (eds) (2014), *Foucault Handbuch. Leben – Werk – Wirkung. Sonderausgabe*, Verlag J. B. Metzler.

Kumar, A., Diksha, A. C. Pandey and M. L. Khan (2020), 'Urban Risk and Resilience to Climate Change and Natural Hazards: A Perspective from Million-Plus Cities on the Indian Subcontinent', in P. K. Srivastava, S. K. Singh, U. C. Mohanty and T. Murty (eds), *Techniques for Disaster Risk Management and Mitigation*, 33–46, John Wiley & Sons.

Kumar, H. (2019), '32 Dead as Worst Flooding in a Decade Hits Booming Mumbai', *New York Times*, 2 July. Available at: https://www.nytimes.com/2019/07/02/world/asia/32-dead-as-worst-flooding-in-a-decade-hits-booming-mumbai.html (accessed 6 November 2024).

Kurtz, H. E. (2003), 'Scale Frames and Counter-Scale Frames: Constructing the Problem of Environmental Injustice', *Political Geography* 22 (8): 887–916. https://doi.org/10.1016/j.polgeo.2003.09.001.

Lawrence, J. L. (2018), 'Manufacturing Biopolitical Disaster: Instrumental (Ir)rationality and the Deepwater Horizon Disaster', in J. L. Lawrence and S. M. Wiebe (eds), *Biopolitical Disaster* (Interventions), 15–29, Routledge/Taylor & Francis Group.

MacKinnon, D. (2010), 'Reconstructing Scale: Towards a New Scalar Politics', *Progress in Human Geography* 35 (1): 21–36. https://doi.org/10.1177/0309132510367841.

Marchezini, V. (2015), 'The Biopolitics of Disaster: Power, Discourses, and Practices', *Human Organization* 74 (4): 362–71. https://doi.org/10.17730/0018-7259-74.4.362.

Marks, D. (2015), 'The Urban Political Ecology of the 2011 Floods in Bangkok: The Creation of Uneven Vulnerabilities', *Pacific Affairs* 88 (3): 623–51. https://doi.org/10.5509/2015883623.

Marks, D. and L. Lebel (2016), 'Disaster Governance and the Scalar Politics of Incomplete Decentralization: Fragmented and Contested Responses to the 2011 Floods in Central Thailand', *Habitat International* 52:57–66. https://doi.org/10.1016/j.habitatint.2015.08.024.

Marquardt, N. and V. Schreiber (2021), '8 Wenn Raumproduktionen zu Regierungspraktiken werden – Michel Foucaults Angebote an die Geographie', in G. Glasze (ed.), *Handbuch Diskurs und Raum. Theorien und Methoden für die Humangeographie sowie die sozial- und kulturwissenschaftliche Raumforschung. Unter Mitarbeit von Annika Mattissek* (Sozial- und Kulturgeographie), 3rd edn, vol. 11, 195–206, transcript verlag.

Marston, S. A. (2000), 'The Social Construction of Scale', *Progress in Human Geography* 24 (2): 219–42.

Mattissek, A. (2010), 'Stadtmarketing in der neoliberalen Stadt. Potentiale von Gouvernementalitäts- und Diskursanalyse für die Untersuchung aktueller Prozesse der Stadtentwicklung', in J. Angermüller and S. van Dyk (eds), *Diskursanalyse meets Gouvernementalitätsforschung. Perspektiven auf das Verhältnis von Subjekt, Sprache, Macht und Wissen* (Sozialwissenschaften), 129–54, Campus-Verlag.

Meerow, S. and C. L. Mitchell (2017), 'Weathering the Storm: The Politics of Urban Climate Change Adaptation Planning', *Environment and Planning A: Economy and Space* 49 (11): 2619–27.

Mehta, L., S. Srivastava, H. N. Adam, B. Alankar, G. Shibaji, K. Upasona and V. Vijay (2019), 'Climate Change and Uncertainty from "Above" and "Below": Perspectives from India', *Regional Environmental Change* 19:1533–47.

Methmann, C. P. (2010), '"Climate Protection" as Empty Signifier: A Discourse Theoretical Perspective on Climate Mainstreaming in World Politics', *Millennium* 39 (2): 345–72. https://doi.org/10.1177/0305829810383606.

Mulye, A. (2018), 'Mumbai's New Development Plan Opens Up Locked Lands, But Will It Be Successful?', *New Indian Express*, 20 November. Available at: https://www.newindianexpress.com/cities/mumbai/2018/Nov/20/mumbais-new-development-plan-opens-up-locked-lands-but-will-it-be-successful-1900708.html (accessed 6 December 2024).

Mumbai March (2020), 'Making of Check Dams'. Available at: https://www.mumbaimarch.com/copy-of-meeting-with-shri-popatrao- (accessed 9 September 2024).

Murali, R. M., M. J. Riyas, K. N. Reshma and S. S. Kumar (2020), 'Climate Change Impact and Vulnerability Assessment of Mumbai City, India', *Natural Hazards* 102 (2): 575–89. https://doi.org/10.1007/s11069-019-03766-2.

Nygren, A. (2016), 'Socially Differentiated Urban Flood Governance in Mexico: Ambiguous Negotiations and Fragmented Contestations', *Journal of Latin American Studies* 48 (2): 335–65. https://doi.org/10.1017/S0022216X15001170.

O'Grady, N. (2018), *Governing Future Emergencies: Lived Relations to Risk in the UK Fire and Rescue Service*, Palgrave Macmillan. https://doi.org/10.1007/978-3-319-71991-7.

O'Grady, N. and D. Shaw (2023), 'Resilience, Responsibility and State Abandon: The Changing Role of the Government in Emergencies', *Political Geography* 100:102796. https://doi.org/10.1016/j.polgeo.2022.102796.

O'Grady, N., D. Shaw and S. Parzniewski (2022), 'People in a Pandemic: Rethinking the Role of "Community" in Community Resilience Practices', *Geoforum* 132:32–41. https://doi.org/10.1016/j.geoforum.2022.03.021.

O'Malley, P. (2009), 'Governmentality and Risk', Legal Studies Research Paper No. 09/98, Sydney Law School. Available at: https://papers.ssrn.com/sol3/papers.cfm?abstract_id=1478289 (accessed 28 October 2025).

Oels, A. (2005), 'Rendering Climate Change Governable: From Biopower to Advanced Liberal Government?', *Journal of Environmental Policy & Planning* 7 (3): 185–207. https://doi.org/10.1080/15239080500339661.

Parab-Pandit, S. (2022), 'Mumbai: To Tackle Waterlogging, BMC Installs Automated Flood Gauges', *Free Press Journal*, 25 August. Available at: https://www.freepressjournal.in/mumbai/mumbai-to-tackle-waterlogging-bmc-installs-automated-flood-gauges (accessed 13 September 2024).

Parab-Pandit, S. (2023), 'Mumbai: BMC Increases Capacity of Dadar Underground Water Tank to Prevent Monsoon Flooding', *Free Press Journal*, 22 May. Available at: https://www.freepressjournal.in/mumbai/mumbai-bmc-increases-capacity-of-dadar-underground-water-tank-to-prevent-monsoon-flooding (accessed 9 September 2024).

Parthasarathy, D. (2009), 'Social and Environmental Insecurities in Mumbai: Towards a Sociological Perspective on Vulnerability', *South African Review of Sociology* 40 (1): 109–26. https://doi.org/10.1080/21528586.2009.10425103.

Parthasarathy, D. (2016), 'Decentralization, Pluralization, Balkanization? Challenges for Disaster Mitigation and Governance in Mumbai', *Habitat International* 52:26–34. https://doi.org/10.1016/j.habitatint.2015.08.022.

Parthasarathy, D. (2021), 'India's Emerging Risk Urbanism', in S. Patel (ed.), *Neoliberalism, Urbanization, and Aspirations in Contemporary India*, 175–94, Oxford University Press.

Peters, G., C. Butsch, F. Krachten, F. Kraas, N. Sridharan and M. A. Marfai (2015), 'Analyzing Risk and Disaster in Megaurban Systems – Experiences from Mumbai and Jakarta', *GRF Davos Planet@Risk* 3 (1): 107–17.

Petzold, J., M. Garschagen, S. Deshpande, R. Dhiman, D. Doshi, A. Katschner, A. Pereira Santos and D. Parthasarathy (2024), 'Identifying Future Challenges for Climate Change Adaptation Through Insights from Participatory Scenario-Downscaling in Mumbai', *npj Climate Action* 3:111. https://doi.org/10.1038/s44168-024-00197-z.

Pinto, R. (2020), 'Mumbai: Half of Flood-Prone Areas "Tackled" But Netas Distrust BMC', *Times of India*, 14 June. Available at: https://timesofindia.indiatimes.com/city/mumbai/mumbai-half-of-flood-prone-areas-tackled-but-netas-distrust-bmc/articleshow/76366286.cms (accessed 13 September 2024).

PNT Bureau (2024), 'Policy Shift in Maharashtra: No-Development Zones Restored, Affordable Housing Plans Altered', *Prop News Time*, 3 January. Available at: https://propnewstime.com/getdetailsStories/NDQxNg==/policy-shift-in-maharashtra-no-development-zones-restored-affordable-housing-plans-altered (accessed 28 October 2025).

The Quint (2022), 'Rising Waters, Sinking City: Mumbai against the Tides'. Available at: https://www.thequint.com/quintlab/mumbai-sinking-city-climate-change-sea-level-rise-documentary/ (accessed 9 September 2024).

Ramalho, J. (2019), 'Worlding Aspirations and Resilient Futures: Framings of Risk and Contemporary City-Making in Metro Cebu, the Philippines', *Asia Pacific Viewpoint* 60 (1): 24–36. https://doi.org/10.1111/apv.12208.

Ramasar, V. (2014), 'Scalar Politics in the South African Waterscape', Doctoral diss., Lund University, Sweden.

Ranganathan, M. and E. Bratman (2021), 'From Urban Resilience to Abolitionist Climate Justice in Washington, DC', *Antipode* 53 (1): 115–37. https://doi.org/10.1111/anti.12555.

Ranger, N., S. Hallegatte, S. Bhattacharya, M. Bachu, S. Priya, K. Dhore, F. Rafique, P. Mathur, N. Naville, F. Henriet, C. Herweijer, S. Pohit and J. Corfee-Morlot (2011), 'An Assessment of the Potential Impact of Climate Change on Flood Risk in Mumbai', *Climatic Change* 104 (1): 139–67. https://doi.org/10.1007/s10584-010-9979-2.

Ravindran, S. (2014), 'Mumbai Will Likely Flood Again – And Nobody's Doing Much About It', *Guardian*, 27 November. Available at: http://www.theguardian.com/cities/2014/nov/27/mumbai-flood-rain-monsoon-city-planning (accessed 13 September 2024).

Rumbach, A. (2016), 'Decentralization and Small Cities: Towards More Effective Urban Disaster Governance?', *Habitat International* 52:35–42. https://doi.org/10.1016/j.habitatint.2015.08.026.

Rutland, T. and A. Aylett (2008), 'The Work of Policy: Actor Networks, Governmentality, and Local Action on Climate Change in Portland, Oregon', *Environment and Planning D: Society and Space* 26 (4): 627–46. https://doi.org/10.1068/d6907.

Sharma, U., A. Scolobig and A. Patt (2012), 'The Effects of Decentralization on the Production and Use of Risk Assessment: Insights from Landslide Management in India and Italy', *Natural Hazards* 64 (2): 1357–71. https://doi.org/10.1007/s11069-012-0300-8.

Shaw, R. (2012), 'Overview of Community-Based Disaster Risk Reduction', in J. M. Pulhin, J. J. Pereira and R. Shaw (eds), *Climate Change Adaptation and Disaster Risk*

Management: An Asian Perspective (Community, Environment and Disaster Risk Management), Bd. 10, vol. 5, 3–17, Emerald.

Shaw, R. (2016), 'Community-Based Disaster Risk Reduction', in R. Shaw (ed.), *Oxford Research Encyclopedia of Natural Hazard Science*, 15–27, Oxford University Press.

Sherly, M. A., S. Karmakar, D. Parthasarathy, T. Chan and C. Rau (2015), 'Disaster Vulnerability Mapping for a Densely Populated Coastal Urban Area: An Application to Mumbai, India', *Annals of the Association of American Geographers* 105 (6): 1198–220. https://doi.org/10.1080/00045608.2015.1072792.

Singh, L. (2020), 'Taking Cue from Tokyo, South Korea, BMC Mulls Underground Ponds to Store Excess Rainwater, Mitigate Flooding', *Indian Express*, 9 August. Available at: https://indianexpress.com/article/cities/mumbai/bmc-mulls-underground-ponds-to-store-excess-rainwater-mitigate-flooding-6546539/ (accessed 9 September 2024).

Singh, L. (2021), 'Mumbai: City's First Underground Water Storage Tanks to Come Up Near Hindmata', *Indian Express*, 10 March. Available at: https://indianexpress.com/article/cities/mumbai/mumbai-citys-first-underground-water-storage-tanks-to-come-up-near-hindmata-7221606/ (accessed 9 September 2024).

Smith, N. (1993), 'Homeless/Global: Scaling Places', in J. Bird, B. Curtis, T. Putnam and L. Tickner (eds), *Mapping the Futures: Local Cultures, Global Change* (Futures, New Perspectives for Cultural Analysis), 87–120, Routledge.

Solecki, W., M. Pelling and M. Garschagen (2017), 'Transitions Between Risk Management Regimes in Cities', *Ecology and Society* 22 (2): 38. https://doi.org/10.5751/ES-09102-220238.

STUP Consultants and Ernst & Young (2016), 'Environmental Impact Assessment Report', Consultancy Services for Preparation of Feasibility Report, DPR Preparation, Report on Environmental Studies and Obtaining MOEF Clearance and BID Process Management for Mumbai Coastal Road Project. Hg. v. Municipal Corporation of Greater Mumbai. Available at: https://portal.mcgm.gov.in/irj/go/km/docs/documents/Coastal%20road/Coastal%20Road%20DPR/EIA%20report.pdf (accessed 4 September 2024).

Surve, S. (2022), 'Mumbai Rains: BMC Now Eyes Footpaths to Tackle Chronic Flooding', *mid-day*, 16 June. Available at: https://www.mid-day.com/mumbai/mumbai-news/article/mumbai-rains-bmc-now-eyes-footpaths-to-tackle-chronic-flooding-23231802 (accessed 13 September 2024).

Swyngedouw, E. (2004), 'Scaled Geographies: Nature, Place, and the Politics of Scale', in E. S. Sheppard and R. B. McMaster (eds), *Scale and Geographic Inquiry: Nature, Society, and Method*, 129–53, Blackwell.

Tapestry Project (2024), 'About TAPESTRY'. Available at: https://tapestry-project.org/about/ (accessed 9 September 2024).

TERI (2014), 'Assessing Climate Change Vulnerability and Adaptation Strategies for Maharashtra: Maharashtra State Adaptation Action Plan on Climate Change (MSAAPC)', The Energy and Resources Institute (TERI), Prepared for the Department of Environment, Government of Maharashtra.

Tozer, L. (2018), 'Urban Climate Change and Sustainability Planning: An Analysis of Sustainability and Climate Change Discourses in Local Government Plans in Canada', *Journal of Environmental Planning and Management* 61 (1): 176–94. https://doi.org/10.1080/09640568.2017.1297699.

UN-HABITAT (2012), *The State of Asian Cities 2010/11*, Fukuoka.

UNISDR (2007), 'Hyogo Framework for Action 2005–2015: Building the Resilience of Nations and Communities to Disasters', Extract from the Final Report of the World Conference on Disaster Reduction (A/CONF.206/6). Hg. v. United Nations International Strategy for Disaster Reduction. Available at: https://www.preventionweb.net/files/1037_hyogoframeworkforactionenglish.pdf (accessed 28 October 2025).

UNISDR (2009), 'Building a Local Government Alliance for Disaster Risk Reduction "The Incheon Declaration"', in Summary from 11–13 August 2009 Conference (After Comments and Approval in Plenary), 1–11, United Nations Office for Disaster Risk Reduction, Incheon.

UNISDR (2015), 'Sendai Framework for Disaster Risk Reduction 2015–2030', United Nations Office for Disaster Risk Reduction, Geneva. Available at: https://www.undrr.org/publication/sendai-framework-disaster-risk-reduction-2015-2030 (accessed 27 August 2021).

Vale, L. J. (2014), 'The Politics of Resilient Cities: Whose Resilience and Whose City?', *Building Research & Information* 42 (2): 191–201. https://doi.org/10.1080/09613218.2014.850602.

van Riet, G. (2016), *The Institutionalisation of Disaster Risk Reduction: South Africa and Neoliberal Governmentality*, Routledge. Available at: https://www.taylorfrancis.com/books/mono/10.4324/9781315464091/institutionalisation-disaster-risk-reduction-gideon-van-riet (accessed 28 October 2025).

Wakefield, S. (2020), 'Urban Resilience as Critique: Problematizing Infrastructure in Post-Sandy New York City', *Political Geography* 79:102148. https://doi.org/10.1016/j.polgeo.2020.102148.

Weinstein, L., A. Rumbach and S. Sinha (2019), 'Resilient Growth: Fantasy Plans and Unplanned Developments in India's Flood-Prone Coastal Cities', *International Journal of Urban and Regional Research* 43 (2): 273–91. https://doi.org/10.1111/1468-2427.12743.

Wissen, M. (2007), 'Politics of Scale: Multi-level-Governance aus der Perspektive kritischer (Raum-)Theorien', in A. Brunnengräber and H. Walk (eds), *Multi-level-*

Governance. Klima-, Umwelt- und Sozialpolitik in einer interdependenten Welt (Schriften zur Governance-Forschung), vol. 9, 229–50, Nomos.

Wolff, E. (2021), 'The Promise of a "People-Centred" Approach to Floods: Types of Participation in the Global Literature of Citizen Science and Community-Based Flood Risk Reduction in the Context of the Sendai Framework', *Progress in Disaster Science* 10:100171. https://doi.org/10.1016/j.pdisas.2021.100171.

Zeiderman, A. (2012), 'On Shaky Ground: The Making of Risk in Bogotá', *Environment and Planning A: Economy and Space* 44 (7): 1570–88.

Zeiderman, A. (2013), 'Living Dangerously: Biopolitics and Urban Citizenship in Bogotá, Colombia', *American Ethnologist* 40 (1): 71–87. https://doi.org/10.1111/amet.12006.

Ziervogel, G., M. Pelling, A. Cartwright, E. Chu, T. Deshpande, L. Harris, K. Hyams, J. Kaunda, B. Klaus, K. Michael, L. Pasquini, R. Pharoah, L. Rodina, D. Scott and P. Zweig (2017), 'Inserting Rights and Justice into Urban Resilience: A Focus on Everyday Risk', *Environment and Urbanization* 29 (1): 123–38. https://doi.org/10.1177/0956247816686905.

Zimmermann, T. (2019), 'Exceeding the Imaginable – Changing Perspectives on Monsoon in Mumbai', in L. Bremner (ed.), *Monsoon [+ Other] Waters*, 235–44, A Monsoon Assemblages Publication.

Zimmermann, T., S. Shinde, D. Parthasarathy and N. C. Narayanan (2023), 'Linking Climate Change Adaptation and Disaster Risk Reduction: Reconceptualizing Flood Risk Governance in Mumbai', *Journal of Integrative Environmental Sciences* 20 (1): 1–29. https://doi.org/10.1080/1943815X.2023.2169712.

Conclusion

Human Geographies of Disaster and Our Current Conjuncture: Taking Stock, Moving Forward

Nathaniel O'Grady and Gemma Sou

From hurricanes and earthquakes to landslides, flooding and volcanoes, the contributions to this collection have critically engaged with myriad disasters that seriously upend collective life in sites across the world. To do so, however, our contributors draw on geographical thought to treat disasters differently from conventional approaches. Throughout the collection, disasters have not been conceptualized in a way that fits their orthodox rendering as exceptional space-times insulated from the broader realities in which they are situated. Rather, disasters have been considered as deeply entwined with the long-standing and specific processes through which different places take shape. Careful consideration of how place shapes and is shaped by disasters is one of the key benefits of geographical thought, and each chapter demonstrates this across diverse contexts.

There are discrete dynamics at play in the relationship between disasters and place-making, and human geography is well placed to reveal them. Disasters are deeply intertwined with the processes through which places are constructed, experienced and understood. On one hand, the way disasters unfold reflects how they become enveloped in, and underpinned by, the unique place-making processes of different locales. On the other, however, disasters are powerful agents of change, actively reshaping the contours of place and the effects that place bears on collective life across multiple scales. This relationship determines how disasters are perceived and remembered, embedding them within the cultural, social and spatial narratives of a place. In this vein, the collection demonstrated how disasters are ingrained within caste systems and at the same time influence relations between peoples of different castes. It brought to life how disasters figure within, but also dictate, the content of ledger lines. It also reflected on how disasters are enrolled into urbanization while actively influencing the forms that cities take. It even documented how disasters are familiarized as kin despite the oft-times existential threat they pose.

In this brief conclusion, we want to continue with this specific mode of critique to show how the collection demonstrates the generative possibilities of human geography for understanding disasters and how it pushes our understanding of disasters in new ways. We do so by asking questions which explore the relationship between disasters and their concurrent conditions along new trajectories. Beyond noting the reciprocity of disasters and wider conditions, we ask: What might the litany of disasters discussed in the collection indicate about the broader historical moment to which we are contemporary? How, in other words, do disasters render distinct the historico-political environments of today? Second, we seek to connect these reflections on the present, and the disasters unfolding therein, with what might come in the future. Rather than making projections, however, we purposefully seek to intervene in the making of this future by reflecting on the contributions in this collection, asking: How might the sense of the contemporary that disasters imply serve as the basis for future research into disasters and beyond?

An Agenda for Future Geographical Engagement with Disasters

Disasters and Our Present Conjuncture

Recently, across the social sciences, the notion of conjuncture has undergone a revival owing to its utility for interpreting the significance that a range of events and processes bear for our broader sense of a shared present. Conjuncture, of course, has come to mean many things simultaneously. And, as such, it draws out the perceived gravitas of our contemporary moment in different ways. Perhaps most obviously, conjunctures emphasize how our current moment is produced at the intersection of multiple, sometimes contradictory phenomena, trends and tendencies that overlap and interact – what Stuart Hall called the 'condensation of contradictions' that give a period its distinctive shape. Our sense of the contemporary is thus formed through the relations between these different occurrences rather than by any single dominant process. As such, the notion of conjuncture leads us to inquire into the associations that disasters bring into play and the ramifications these connections hold for our understanding of the present, opening up the possibility for alternative associations and new political or social imaginaries to emerge.

Read in such a way, the chapters in the collection might be interpreted as brimming with an even deeper sense of foreboding than first thought. Such is the case, for one, because of the importance of climate change in shaping our understanding of the here and now. For Timothy Morton (2013), climate change is a hyper-object – a condition, process or event so ubiquitous and significant that it can only be considered in the abstract. It haunts us, being both present and absent at one and the same time. Thought of conjuncturally in relation to disasters, however, perhaps we can backtrack from Morton's argument somewhat. This is because many of the disasters discussed in the collection could be considered to function as iterations of a changing climate. Disasters, in many ways, operate as records of the ways in which the broader climate is experiencing turbulence. They thus take the abstraction of climate change and render it specific and material. As the climate crisis worsens, the frequency and severity of many of these disasters are intensifying and accelerating. While warmer ocean temperatures exacerbate hurricane risk, changing wind patterns and neglected fuel loads are causing a surge in wildfires. These wildfires interact with eroding coastlines to increase landslides and flooding as well. At perhaps a more rudimentary level, temperatures have soared over the last century, meaning extreme heat is becoming an ever more dangerous occurrence in the summer months.

Climate change, like the disasters that often serve as its iterations, is not autonomous from other prevailing forces that make up our current moment. Instead, it intersects with shifts in emerging political realities across the world, which are now as unsettled as they have ever been since the Second World War. Over the past decade, historical orders that have, inadequately, been described as 'neoliberal' have, in many ways, disassembled before our eyes. One of the most notable replacements has been a surge in authoritarian tendencies in countries across the world. Indicators of authoritarianism manifest in different forms. Borders that were open during the large passages of both the twentieth and twenty-first centuries, as part of the neoliberal project, are beginning to close for many. Similarly, markets that were to a degree unregulated and free are the object of new forms of confinement and restriction. These indicators must be conceived as epistemological too – signs of deeper transformations in how societies organize and perceive themselves. A variety of concerning modes of relation to climate change are at the heart of emerging authoritarian projects. Climate scepticism, for instance, has become a prominent attitude in our post-neoliberal world. Here, the environmental turbulence of today is compared to previous centuries and millennia, working to dampen calls for urgent action. In other cases, climate action is represented as a threat to established ways of life, whether in the Global North or in the Global South that is still 'developing'. In some instances, continuing with the use of carbon-intensive industries is even celebrated, often in a way underscored with obvious ecocidal intent.

Through allusions to 'greenwashing' and the like, a cadre of literature has argued that government-orchestrated climate adaptation and disaster risk reduction can reproduce the status quo by stripping disasters of their potential to instigate widespread change. In this view, such measures, often associated with the liberal project, are seen as mechanisms that absorb the impetus for deeper political change that might stem from climate change, channelling it into technocratic management rather than systemic transformation. However, these putative traits of the liberal project might be said to pale in significance when contrasted with emerging authoritarian regimes in which failure to act on climate change, and sometimes an active celebration of environmental degradation, is expressed as a core ideological commitment.

Geographical Thought as Alternative Critique of Disaster

The rising tide of authoritarianism and its association with climate scepticism thus reveals a gap in our critical architecture in which allusions to liberal co-optation no longer hold as much sway as before. Rather than seeking to dominate the narrative in dealing with a reality upon which we can all agree, the emerging political conjuncture is characterized by conflicts over reality itself (Latour 2018). This gap leads us to our second question: How might this collection contribute to the role of scholarship in responding to our present?

Conjunctural analysis is not only about connecting different events and processes. In the Great Moving Right Show, Stuart Hall (2019) uses conjunctural analysis to advocate for more critically reflexive modes of scholarship. It is insufficient, then, to denounce developments in the world and make grave projections. Instead, conjunctural analysis encourages us to reflect on the inadequacies of our own responses to crisis. In many ways, the contributions to this collection exemplify how conjunctural analysis can address the crisis of imagination that has permeated academia in recent years. In so doing, they offer valuable directions for future research agendas in disaster geographies and in geography more widely.

One agenda would concern the forms of critique that might orient the future of disaster geographies. The analyses found in the collection are conscious attempts to avoid the kind of reductionism that has characterized recent scholarship. Too often, 'critical' is considered to amount to parcelling phenomena into broad structural categories such as 'neoliberal' or 'colonial' and so on. While perhaps retaining importance owing to their capacity to connect different scales of collective life, such critiques risk limiting and confining the forms of analysis that we deploy to make sense of the world. By moving beyond structural analyses, the chapters in this collection deploy a different kind of critical thinking. It is one perhaps more closely aligned to the spirit, if not the content, that characterized critical theory as it existed after the Second World War, exemplified by Adorno and Horkheimer's Dialectic of Enlightenment (1997). Here, critique exceeds denunciation and concerns itself with both identifying key ruptures in the forms of reasoning which come to organize societies and developing new forms of analysis to address them. Critique in the chapters thus inspires creative and innovative ways of thinking. It is only by mobilizing such renditions of critique that shift and expand our understanding of the present that we can hope to address the complexities of our current moment.

Moving away from structural-denunciatory critique is also of importance for the methodological agenda of this collection. As Clive Barnett (2017) reminds us, relying on prescribed structural categories often leads to deductive inquiry, in which investigations of the world become equivalent to gauging the extent to which our research might be constrained within existing knowledge and theory (see O'Grady (2025) for a discussion in relation to disaster resilience). In contrast, the contributions to this collection call for more inductive modes of inquiry that seek to understand their cases on their own terms. This shift is vital if we are to keep pace with the ebb and flow of the processes that constitute the contemporary. Inductive methodologies foster an openness in our approach, enabling us to identify new problems as they emerge. In turn, responses that are unbound from existing trends, yet well calibrated to such problems, can be developed.

A concern with and for the empirical is central to our call for more inductive modes of inquiry. The epistemological turn towards scepticism that characterizes authoritarianism will surely gather pace and weight if it is not countered by an insistence on the evidentiary character of the claims we make. Such a sentiment is unfortunately lacking in much recent literature on disasters and climate change. However, the creative analyses and inductive methodologies showcased in this collection are marked by a commitment to comprehensive and nuanced accounts of diverse cases and field sites. Rather than skating over nuance and complexity for the sake of spectacular, seductive and simplistic arguments or calls to arms, the chapters profess a commitment to communicating the messy reality of life. The decision to stay with the always complicated character of disasters is an ethical one, guided by a belief that glossing over the intricacies of disaster would risk reducing scholarship to the realm of sensationalism and reactionary 'hot takes' that too often dominate public discourse today.

Conclusion

We hope that the insights and ideas showcased in this collection on the geographies of disasters will be given their due place in academic discourse and receive the scholarly attention that we argue they deserve. One of our central aims has been to demonstrate how human geography opens up new and innovative pathways for researching disasters, moving beyond traditional approaches to reveal the complex place-based processes that shape both the causes and consequences of disasters. By drawing upon a diverse set of research, we set out to illustrate how bridging human geography and disaster studies is not only possible but also highly rewarding, offering much-needed refinement to the ways in which disasters are conceptualized, studied and addressed. Although not explicitly discussed in the chapters, we foresee that the collection will serve as a useful teaching resource. Students will appreciate the place-based approaches across the collection, which illustrate and make accessible seemingly arcane conceptual issues such as 'disaster governance', 'vulnerability' or 'recovery'. For this reason, we can hope that both students and researchers alike may be inspired by the work within this collection to pursue similar geographical inroads into the study of disaster.

Looking ahead, the future of human geographical research on disasters will be defined by an increased focus on the interconnectedness of risks and the need for imaginative and transformative solutions. As climate change, rapid urbanization and resource depletion intensify the frequency and complexity of disasters, we argue that human geographers are well placed to play a crucial role in mapping out rigorous, place-based responses that address not only immediate hazards but also the underlying societal structures and inequalities that perpetuate vulnerability. At the same time, human geography will continue to challenge surface-level fixes by questioning prevailing mindsets and advocating for deep systemic change, reimagining responsibility, value and our relationship with the environment. For this reason, we argue that human geographical research will remain at the forefront of studying disasters, shaping more equitable, just and sustainable futures.

References

Adorno, T. W. and M. Horkheimer (1997), *Dialectic of Enlightenment*, Verso.

Barnett, C. (2017), *The Priority of Injustice: Locating Democracy in Critical Theory*, vol. 36, University of Georgia Press.

Hall, S. (2019), *Essential Essays*, vol. 1, Duke University Press.

Latour, B. (2018), *Down to Earth: Politics in the New Climatic Regime*, John Wiley & Sons.

Morton, T. (2013), *Hyperobjects: Philosophy and Ecology After the End of the World*, University of Minnesota Press.

O'Grady, N. (2025), 'On the Possibility of "Just Resilience": A Pragmatist Approach to Justice-Based Climate Change Governance', *Geoforum* 159:104163. https://doi.org/10.1016/j.geoforum.2024.104163.

Index

www.ingramcontent.com/pod-product-compliance
Ingram Content Group UK Ltd.
Pitfield, Milton Keynes, MK11 3LW, UK
UKHW020214290926
15470UKWH00001B/170

9 781666 97088